Study Guide

Algebra 2

Explorations and Applications

This Study Guide includes a section for every section in the textbook. Each Study Guide section contains an illustrated list of Terms to Know, worked-out Examples, and practice exercises, including Spiral Review questions. Each Study Guide chapter contains a Chapter Review consisting of a Chapter Check-Up and a Spiral Review of all previous chapters. Answers are provided in a separate Answer Key.

McDougal Littell

A Houghton Mifflin Company

Evanston, Illinois

Boston   Dallas

ISBN: 0-395-76964-7

123456789 - HS - 00 99 98 97 96

CONTENTS

Studying Algebra 2···············

Think about how you learn new material in mathematics. Do you learn best

- **by reading?**
- **by listening?**
- **by writing things down?**
- **by explaining to someone else?**
- **or by some combination of these methods?**

Thinking about what works best for you can help you make the best use of your study time and to remember what you have learned.

Students who learn best by *thinking things through* alone often help themselves remember by putting their thoughts in writing. Making a concept map or writing a paragraph that summarizes the material can help you think about how the information is organized.

Students who learn best by *talking things through* can discuss the material with another student or with a study group. By sharing ideas, students can often help each other learn and remember mathematical concepts.

Study Tips

Here is a list of some activities that may help you learn new material, or review material that you learned previously. You should decide whether each of these is a type of activity that has helped you learn things in the past or that might help you learn things in the future.

1. Reading

- Read the "Learn how to . . ." information at the beginning of each section and see if you can predict the types of problems you will be asked to do.

- Read the section the night *before* it is presented in class. This can help by focusing your attention on the points that you did not understand when you read about them.

- Reread the section after it is assigned, looking for any concepts that are still unclear.

- Reread any notes that you have written and see if you remember what you were thinking about when you wrote them.

2. Reading and Writing

- Read and outline the ideas in the section or chapter.

- Read each worked-out example and think about the concepts involved; then copy the problem and try to solve it without referring to the solution.

- Read each worked-out example and then write a similar problem to solve on your own.

---

3. Writing

- Make an outline of the section or chapter.

- Make a concept map for each section.

- Redo any problem you found difficult, comparing your new work with your previous work.

- Write a paragraph that summarizes the section or chapter.

- List the key words from the section and write their definitions in your own words.

4. Explaining

- Compare your ideas with those of several other students.

- Form a study group to work together and to help each other.

- Discuss and compare your concept maps for each chapter with other students.

- Identify a few exercises from each section in the chapter that illustrate the main objective of the section.

- Discuss the study techniques that have helped you most in the past with other students.

Mathematics Journals

Writing in a journal helps you keep a record of the things you learn about mathematics, as well as the things you learn about *learning* mathematics. When you write a paragraph, you organize your thoughts and record the important ideas that you have learned.

Here are some suggestions for writing in your journal that will help you study new material or review material that you learned earlier.

- Write about your favorite topic in mathematics and how you learned this topic.

- Write a description of the technique that helps you most when you learn new concepts. Explain why you think this helps you learn.

- Write a description of the technique that helps you most when you review a chapter. Did you *write* a summary or list of topics, or did you summarize by *talking* to someone else in your class?

- Look back at your work in previous chapters to see if your technique has changed. If it has changed, how has it changed and why do you think it changed? If it has not changed, explain why this technique seems to be working for you.

Modeling Growth with Graphs and Tables

Learn how to . . .

- model real-world situations with tables and graphs
- compare different models describing the same situation

So you can . . .

- make predictions about future trends

Application

If you put the same amount of money into the bank every day, the total amount of money in the bank grows at a constant rate.

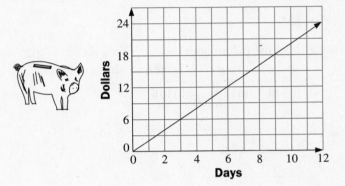

Terms to Know

Mathematical model (p. 4)
  a description of a situation using graphs, tables, or equations

Example / Illustration

A calendar is a mathematical model of the pattern of days in a month.

September 1996

S	M	T	W	Th	F	S
1	2	3	4	5	6	7
8	9	10	11	12	13	14
15	16	17	18	19	20	21
22	23	24	25	26	27	28
29	30					

UNDERSTANDING THE MAIN IDEAS

Constant growth models

A constant growth model assumes that a quantity will increase (or decrease) by about the same amount over equal periods of time.

Example 1

Janet Dole kept a record of her baby daughter Julia's weight during her first nine months, as shown in the table at the right. Use a constant growth model to estimate Julia's weight on her first birthday.

Age (mo)	Weight (lb)
0	8.7
3	13.7
6	18.2
9	21.4

▬ Solution ▬

Step 1 Use the information about Julia's weight to determine how much weight she gained during each three-month time period.

Age (mo)	Weight (lb)
0	8.7
3	13.7
6	18.2
9	21.4

> 5.0 lb
> 4.5 lb
> 3.2 lb

Note: The measurements must occur at equally-spaced intervals; in this case, the baby's weight was recorded every three months.

Step 2 Find the average (mean) of the increases in weight.

$$\frac{5.0 + 4.5 + 3.2}{3} \approx 4.2 \text{ lb}$$

Step 3 Use the average increase to estimate Julia's weight at 1 yr (12 mo).

Since Julia weighed 21.4 lb at 9 mo, her weight three months later will be about 21.4 + 4.2, or 25.6 lb.

A constant growth model predicts that Julia will weigh about 25.6 lb on her first birthday.

---

1. Suppose a baby weighs 8.7 lb at birth but, unlike Julia, gains exactly 4.2 lb every three months. Make a table showing this baby's weights at 0, 3, 6, 9, and 12 mo.

2. Complete this histogram comparing baby Julia's actual weights with the weights found in Exercise 1.

3. Writing How does the constant growth model compare with the actual growth data?

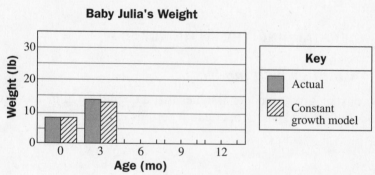

Baby Julia's Weight

Key
■ Actual
▨ Constant growth model

---

4. Use the constant growth model to predict Julia's weight at age 2 and at age 10. Do you think this is a valid model to predict Julia's weight in her teens?

The table at the right shows the monthly attendance at a club's meetings.

5. Find the change in attendance for each pair of successive months. Then find the average (mean) monthly change in attendance. (*Hint:* Your answers should be negative.)

6. Use the the average change in attendance found in Exercise 5 to predict the attendance at the club's January meeting.

Month	Number attending
September	206
October	175
November	142
December	107

Proportional growth model

A proportional growth model assumes that a quantity will increase (or decrease) about the same percent over equal periods of time.

Example 2

Use a proportional growth model to estimate baby Julia's weight at 12 mo.

■ Solution ■

Step 1 Find the average (mean) of the percent increases in Julia's weight over each 3-month period.

Age (mo)	Weight (lb)	Increase from last weighing	Percent increase
0	8.7	–	–
3	13.7	5.0	$\frac{5.0}{8.7} \approx 57.5\%$
6	18.2	4.5	$\frac{4.5}{13.7} \approx 32.8\%$
9	21.4	3.2	$\frac{3.2}{18.2} \approx 17.6\%$

Note: When finding the percent increase, you must divide the increase in weight by the *previous* weight.

The average (mean) of the percent increases is $\frac{57.5 + 32.8 + 17.6}{3}$, or about

36.0%. So, the model predicts that Julia's weight at 12 mo will be a approximately 36.0% greater than her weight at 9 mo.

(Solution continues on next page.)

Step 2 Find the *growth factor*. The growth factor is the decimal equivalent of the sum of 100% and the average percent increase.

100% + 36.0% = 136.0%, or 1.36

Step 3 Use the growth factor to predict Julia's weight at 12 mo.

$$\begin{matrix}\text{growth} \\ \text{factor}\end{matrix} \times \begin{matrix}\text{previous} \\ \text{weight}\end{matrix} = \begin{matrix}\text{predicted} \\ \text{weight}\end{matrix}$$

1.36 × 21.4 ≈ 29.1 lb ← The previous weight is her 9 mo weight.

A proportional growth model predicts that Julia will weigh about 29.1 lb on her first birthday.

7. Make a table, similar to the one you made for Exercise 1, showing what a baby weighing 8.7 lb at birth would weigh if its weight increased exactly 36% every three months.

8. Make a histogram, similar to the one you completed for Exercise 2, comparing baby Julia's actual weights to those in your table for Exercise 7.

9. **Writing** Baby Julia actually weighed 24.4 lb at 12 mo. In fact, she did not reach 29.1 lb until she was 21 mo old. Why do you think the proportional growth model is a poor model for this situation?

For Exercises 10–12, refer to the club attendance data given in the table for Exercises 5 and 6.

10. Find the percent change in attendance for each month. Then find the average (mean) percent change in attendance. (*Hint*: Should your answers be positive or negative?)

11. What is the corresponding "growth" factor for the average percent change found in Exercise 10?

12. Use the "growth" factor from Exercise 11 to predict the club's attendance at their January meeting.

13. Make a histogram to compare the club's actual attendance data with the attendance data generated by a constant growth model (Exercises 5 and 6) and the attendance data generated by a proportional growth model (Exercises 10–12).

14. **Open-ended** Which model better represents the attendance data? Explain.

Protein and Calorie Content in an Average Serving		
Food	Calories	Protein (g)
1 cup milk (2%)	145	10
1 cup rice	225	4
1 medium stalk broccoli	50	6
1/2 cup cooked red beans	150	20
1 medium banana	100	1

15. Make a scatter plot using the calorie and protein data. (*Toolbox, page 795*)

For each equation, make a table of values for *x* and *y* when *x* = 0, 1, 2, and 3.
(*Toolbox, page 795*)

16. $y = 3x - 20$ **17.** $y = 5(3^x)$

Simplify. Give answers with the appropriate number of significant digits.
(*Toolbox, page 789*)

18. $60.05 + 3.027$ **19.** 35.6×3.005

1.2

Using Functions to Model Growth

Learn how to . . .

- write a function to model a set of data

So you can . . .

- make predictions about a situation

Application

The function $C = 25 + 30h$ models the cost C of a service call lasting h hours.

Service Call Charges
$25 plus
$30 per hour

Terms to Know

Example / Illustration

Terms to Know	Example / Illustration
Function (p. 10) a table or rule that pairs each input value with exactly one output value	$C = 25 + 30h$
Independent variable (p. 10) the variable that represents the input of a function	In the function above, the independent variable is h. You can substitute any reasonable value for h that you wish.
Dependent variable (p. 10) the variable that represents the output of a function	In the function above, the dependent variable is C. The value of C depends on the value chosen for h, as shown in the table below.

h	C
1	$55
2	$85
3	$115
4	$145

UNDERSTANDING THE MAIN IDEAS

Linear functions

A linear function is a model based on constant growth. Such a function can be written in the form

$$y = b + ax$$

where y is the dependent variable, b is the initial amount, a is the constant growth factor, and x is the independent variable.

Example 1

FunStuff Amusements, a toy store, opened recently. The number of customers at the store has been increasing each week. The manager wrote the linear function $y = 250 + 150x$, where x is the number of weeks that the store has been open, to model the number of customers each week.

a. Which variable is the independent variable? the dependent variable?

b. How many customers shopped at FunStuff the first week it was open?

c. What is the average change in the number of customers each week?

■ Solution ■

a. Since the number of customers depends on the week, x is the independent variable and y is the dependent variable.

b. Using $x = 1$ in the function: $y = 250 + 150(1)$, or $y = 400$. So, 400 customers visited the store during the first week.

c. The constant growth rate, a, in the function is 150. This value represents the average change in the number of customers each week.

---

1. Make a table of values showing the number of customers at the store in each of the first six weeks.

2. Suppose that during the first week there were 300 customers. How would the function change?

3. Suppose that 65 customers visited a store during the week it opened and that each week after that the number of customers was about 50 more than the previous week. Write a linear function modeling this situation.

A video rental store kept a record of the number of times that a particular hit movie was rented each week since it has been available. The data is shown in the table at the right.

Weeks since release	Number of times rented
1	56
2	49
3	41
4	33
5	24

4. Define the independent and the dependent variables in this situation.

5. Find the average (mean) change in the number of rentals of this movie each week.

6. Write a linear function modeling the number of rentals y of this movie during the week x weeks since it became available.

7. Use the model in Exercise 6 to predict how many times the movie will be rented during the week 6 weeks since it became available.

Exponential functions

An exponential function is a model based on proportional growth. Such a function can be written in the form

$$y = a_0(b)^x$$

where y is the dependent variable, a_0 is the initial amount, b is the proportional growth factor, and x is the independent variable.

Example 2

A political pollster took monthly polls to estimate how many voters recognized a certain candidate's name. The pollster's results are given in the table below.

Month	Number of months after January	Estimated number of voters recognizing candidate's name
January	0	7500
February	1	8625
March	2	9930
April	3	11,420

Write an exponential function that models the number of voters y who recognize the candidate's name x months after January.

■ Solution ■

Step 1 Find the average (mean) of the monthly percent increases.

January to February: $\dfrac{8625 - 7500}{7500} = 0.15$, or 15%

February to March: $\dfrac{9930 - 8625}{8625} \approx 0.1513$, or about 15.13%

March to April: $\dfrac{11{,}420 - 9930}{9930} \approx 0.1501$, or about 15.01%

So, the mean monthly percent increase is

$\dfrac{15 + 15.13 + 15.01}{3} \approx 15.05\%$

Step 2 Find the growth factor (expressed as a decimal).

100% + 15.05% = 115.05%, or about 1.15

Step 3 Use the growth factor and the initial amount to write the exponential function

$y = 7500(1.15)^x$

where y is the number of voters recognizing the candidate's name and x is the number of months after January.

---

8. Make a table of values for the function $y = 7500(1.15)^x$ using $x = 0, 1, 2,$ and 3. Compare these values with the actual data in Example 2.

9. Use the exponential model to predict the candidate's name recognition with voters in July and again in October.

While taking a keyboarding course, Marya kept a record of her error rate in mistakes per minute. Her data is shown in the table at the right. Use this information for Exercises 10–13.

Weeks in class	Error rate (mistakes/min)
0	24
1	19
2	15
3	12
4	10

10. Find the average change per week in Marya's error rate, and then find the average percent change in her error rate. (*Note:* Since her error rate is decreasing, your answers should be negative.)

11. Write both a linear function and an exponential function to model Marya's error rate.

12. Use your models from Exercise 11 to predict Marya's error rate after 10 weeks of keyboarding class. Which model do you think gives a better prediction? Why?

13. **Technology** Use a graphing calculator or computer software to graph the linear function and the exponential function you wrote for Exercise 11 on the same set of axes. Which graph is decreasing faster?

For Exercises 14 and 15, write a function to model each situation. Tell whether your function is *linear* or *exponential*.

14. When the new city library opened, there were 40,000 books in its collection. Since then, approximately 1200 books per year have been added to its collection.

15. When the new city library opened, there were 40,000 books in its collection. Since then, the number of books in its collection has been increasing by about 3% per year.

16. **Mathematics Journal** Describe a situation in which an linear growth model would be appropriate. Then describe a situation in which a proportional growth model would be appropriate.

···················
Spiral Review

The table shows the amount, in billions of dollars, that was owed each year in automobile loans in the United States. *(Section 1.1)*

17. Use a constant growth model to estimate the outstanding debt on automobile loans in the year 2000.

18. Repeat Exercise 17 using a proportional growth model.

Year	Estimated total debt (billions of dollars)
1970	36.3
1980	111.9
1990	294.2

For Exercises 19–21, evaluate each expression when *a* = –3 and *b* = 6. *(Toolbox, page 780)*

19. $(a - b)^3$ 20. $-a + b$ 21. $2a - 4b + 16$

22. The table below gives the number of passenger arrivals and departures at five airports in the United States in 1992. Find the mean, median, and mode of the data. *(Toolbox, page 790)*

Airport	Arrivals and departures (billions of passengers)
Logan (Boston)	23
Chicago-O'Hare	64
Dallas-Ft.Worth	52
McCarran (Las Vegas)	21
San Francisco	32

Learn how to . . .

- organize information in a matrix

- add matrices

- multiply a matrix by a scalar

So you can . . .

- use matrices to model data

Modeling with Matrices

Application

You can use scalar multiplication of matrices to model the transformation of triangle ABC into triangle $A'B'C'$ in the figure below.

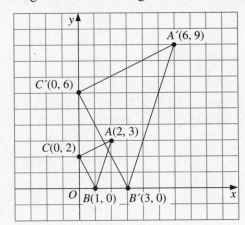

$$3\begin{bmatrix} 2 & 1 & 0 \\ 3 & 0 & 2 \end{bmatrix} = \begin{bmatrix} 6 & 3 & 0 \\ 9 & 0 & 6 \end{bmatrix}$$

Terms to Know	*Example / Illustration*
Matrix (p. 16) a group of numbers arranged in rows and columns (Matrices are labeled using capital letters.)	$A = \begin{bmatrix} 2 & 0 & 8 \\ 3 & -1 & 0 \end{bmatrix}$
Element of a matrix (p. 16) each number in a matrix (You write $a_{1,3}$ to represent the element in row one and column three of a matrix A.)	There are six elements in matrix A above. Element $a_{1,3}$ is 8.
Dimensions of a matrix (p. 16) the form $r \times c$, where r is the number of rows and c is the number of the columns	The dimensions of matrix A above are 2×3.
Scalar (p. 18) a number	In the scalar multiplication shown below, the scalar is 5.
Scalar multiplication (p. 18) the process of multiplying a matrix by a number	$5\begin{bmatrix} -2 & 4 \\ 1 & -3 \end{bmatrix} = \begin{bmatrix} 5(-2) & 5(4) \\ 5(1) & 5(-3) \end{bmatrix}$ $= \begin{bmatrix} -10 & 20 \\ 5 & -15 \end{bmatrix}$

UNDERSTANDING THE MAIN IDEAS

Matrix addition and subtraction

Two matrices can be added or subtracted only if they have the same dimensions. To add two such matrices, you add the corresponding elements—corresponding elements are elements in the same position of their respective matrices. To subtract two matrices with the same dimensions, subtract the corresponding elements. In each case, the result is a matrix with the same dimensions as the original two matrices.

Example 1

On each of two successive weekends, a theater company presented three performances of the play "Our Town." Matrix C below represents the number of children at each performance and matrix A represents the number of adults at each performance.

Matrix C: number of children

$$\begin{array}{c} \\ \text{Friday} \\ \text{Saturday} \\ \text{Sunday} \end{array} \begin{array}{cc} \text{Week 1} & \text{Week 2} \\ \begin{bmatrix} 42 & 36 \\ 29 & 43 \\ 60 & 65 \end{bmatrix} \end{array}$$

Matrix A: number of adults

$$\begin{array}{c} \\ \text{Friday} \\ \text{Saturday} \\ \text{Sunday} \end{array} \begin{array}{cc} \text{Week 1} & \text{Week 2} \\ \begin{bmatrix} 194 & 207 \\ 245 & 226 \\ 193 & 215 \end{bmatrix} \end{array}$$

a. Find $C + A$, and explain what the resulting matrix represents.

b. Find $A - C$, and explain what the resulting matrix represents.

■ Solution ■

a. $C + A = \begin{bmatrix} 42 & 36 \\ 29 & 43 \\ 60 & 65 \end{bmatrix} + \begin{bmatrix} 194 & 207 \\ 245 & 226 \\ 193 & 215 \end{bmatrix} = \begin{bmatrix} 42 + 194 & 36 + 207 \\ 29 + 245 & 43 + 226 \\ 60 + 193 & 65 + 215 \end{bmatrix} = \begin{bmatrix} 236 & 243 \\ 274 & 269 \\ 253 & 280 \end{bmatrix}$

The resulting matrix represents the total attendance at each performance, as shown in the labeled matrix below. (The labels are the same as those for matrices A and C.)

$$\begin{array}{c} \\ \text{Friday} \\ \text{Saturday} \\ \text{Sunday} \end{array} \begin{array}{cc} \text{Week 1} & \text{Week 2} \\ \begin{bmatrix} 236 & 243 \\ 274 & 269 \\ 253 & 280 \end{bmatrix} \end{array}$$

(Solution continues on next page.)

b. $A - C = \begin{bmatrix} 194 & 207 \\ 245 & 226 \\ 193 & 215 \end{bmatrix} - \begin{bmatrix} 42 & 36 \\ 29 & 43 \\ 60 & 65 \end{bmatrix} = \begin{bmatrix} 194-42 & 207-36 \\ 245-29 & 226-43 \\ 193-60 & 215-65 \end{bmatrix} = \begin{bmatrix} 152 & 171 \\ 216 & 183 \\ 133 & 150 \end{bmatrix}$

The resulting matrix shows how many more adults than children attended each performance, as shown in the labeled matrix below.

$$\begin{array}{c} \\ \text{Friday} \\ \text{Saturday} \\ \text{Sunday} \end{array} \begin{array}{cc} \text{Week 1} & \text{Week 2} \\ \begin{bmatrix} 152 & 171 \\ 216 & 183 \\ 133 & 150 \end{bmatrix} \end{array}$$

For Exercises 1–3, refer to the matrices *A* and *C* in Example 1.

1. What is the position of the element "29" in matrix C?

2. What is the element $a_{1,2}$ in matrix A?

3. What are the dimensions of matrices C, A, $A + C$, and $A - C$?

For Exercises 4–6, refer to the egg substitution matrix *E* below. A cook can use this matrix to convert the number of large eggs in a recipe to a number of eggs of another size.

For this number of:		Substitute this number of eggs:			
Large eggs		**Jumbo**	**Extra-Large**	**Medium**	**Small**
1		1	1	1	1
2		2	2	2	3
3	Matrix $E =$	2	3	3	4
4		3	4	5	5
5		4	4	6	7
6		5	5	7	8

4. What are the dimensions of matrix E?

5. What is the position of the element "6" in matrix E? What information does this element represent?

6. In which position can you find the number of medium eggs that can be substituted for 2 large eggs? What is this number?

Scalar multiplication

A matrix can be multiplied by a number using a process called *scalar multiplication*. This process involves multiplying each element of the matrix by the number, called the *scalar*. The scalar multiplication of a matrix D by the number -4 can by written as $-4D$.

Example 2

Children's tickets for the production of "Our Town" in Example 1 sold for $3 each. Write a matrix showing now much money the theater company received for children's tickets at each performance.

■ Solution ■

For each performance, the amount the theater company received for children's tickets was $3 times the number of children's tickets sold. The required matrix is then $3C$.

$$3C = 3\begin{bmatrix} 42 & 36 \\ 29 & 43 \\ 60 & 65 \end{bmatrix} = \begin{bmatrix} 3\cdot 42 & 3\cdot 36 \\ 3\cdot 29 & 3\cdot 43 \\ 3\cdot 60 & 3\cdot 65 \end{bmatrix} = \begin{bmatrix} 126 & 108 \\ 87 & 129 \\ 180 & 195 \end{bmatrix}$$

The resulting matrix shows the amount of money (in dollars) the theater company received from the sale of children's tickets for each performance, as shown in the labeled matrix below.

$$\begin{array}{c} \\ \text{Friday} \\ \text{Saturday} \\ \text{Sunday} \end{array} \begin{array}{cc} \text{Week 1} & \text{Week 2} \end{array} \\ \begin{bmatrix} 126 & 108 \\ 87 & 129 \\ 180 & 195 \end{bmatrix}$$

7. What is the scalar in Example 2?

8. Refer to matrix A in Example 1. Suppose adult's tickets to the theater company's production of "Our Town" sold for $5. Write a matrix that represents the money received from the sale of adult's tickets at each performance.

9. Refer to the matrices in Example 1. Find $3C + 5A$. What does this matrix represent?

For Exercise 10–12, use matrices B and D below to evaluate each matrix expression. If an operation cannot be performed, write undefined.

$$B = \begin{bmatrix} 3 & 0 & 6 \\ -5 & 1 & 8 \end{bmatrix} \qquad D = \begin{bmatrix} 0 & -9 & 0 \\ 4 & 6 & 2 \end{bmatrix}$$

10. $2B - D$

11. $\dfrac{1}{2}B$

12. $B + 0.1D$

Study Guide, ALGEBRA 2: EXPLORATIONS AND APPLICATIONS

The table shows the percentage of households in the United States from 1978 to 1984 having a microwave oven. *(Section 1.2)*

Year	Percentage of homes having a microwave
1978	8%
1980	14%
1982	21%
1984	34%

13. Write a linear function to model the situation.

14. Write an exponential function to model the situation.

15. Estimate the percentage of households in the United States having microwaves in 1986. Explain how you arrived at your estimate.

Simplify. *(Toolbox, page 778)*

16. $-3(-8) - 8(5)$

17. $6 \div 5 + 3 \div 10$

18. $20 \cdot 3 + 7(-6)$

Tell whether each property is true for real numbers *a* and *b*. Give an example to support your answer. *(Toolbox, page 774)*

19. $a \cdot 0 = a$

20. $\frac{b}{a} = \frac{a}{b}, a \neq 0, b \neq 0$

21. $a(b + c) = ab + ac$

22. $a^b = b^a$

23. $(a - b) = -(b - a)$

24. $|a - b| = |a| - |b|$

Multiplying Matrices

Learn how to . . .

- multiply matrices

So you can . . .

- solve problems involving matrix multiplication

Application

You can use matrix multiplication to organize and keep track of multiplication and addition in many categories, like sales records.

Number Sold • Prices = Total Receipts

	Hot Lunch	Snack Pack	Dessert	Milk
Mon	215	84	265	290
Tues	145	153	264	292
Wed	270	15	259	283
Thurs	222	72	243	291
Fri	255	36	261	289

•

	Prices
Hot Lunch	$1.75
Snack Pack	$1.25
Dessert	$.80
Milk	$.35

=

	Receipts
Mon	$794.75
Tues	$758.40
Wed	$797.50
Thurs	$774.75
Fri	$801.20

Terms to Know

Matrix multiplication (p. 23)

the process of multiplying two matrices (For the product of two matrices to be defined, the number of elements in a row of the first (left-hand) matrix must equal the number of elements in a column of the second (right-hand) matrix.)

Example / Illustration

$$\begin{bmatrix} 3 & 0 \\ 4 & 2 \end{bmatrix}\begin{bmatrix} 6 & -8 \\ 0 & -1 \end{bmatrix} =$$

$$\begin{bmatrix} 3(6) + 0(0) & 3(-8) + 0(-1) \\ 4(6) + 2(0) & 4(-8) + 2(-1) \end{bmatrix} =$$

$$\begin{bmatrix} 18 & -24 \\ 24 & -34 \end{bmatrix}$$

UNDERSTANDING THE MAIN IDEAS

Matrix multiplication

You can find the product AB of two matrices A and B only if matrix A has as many columns as matrix B has rows. The product will be a matrix with as many rows as A and as many columns as B. Each element of matrix AB is the sum of the products of each row element of A with its corresponding column element in B. For example, the element in the second row and first column of matrix AB is the sum of the products of each element in the second row of matrix A and its corresponding element in the first column of matrix B.

When multiplying matrices with labels, the column labels of the first matrix must match the row labels of the second matrix (see the Application).

Example

17

A box of low-fat baking mix has several recipes printed on it. Matrix I below shows the ingredients in four of those recipes.

$$\text{matrix } I = \begin{array}{c} \\ \text{Pancakes} \\ \text{Waffles} \\ \text{Biscuits} \\ \text{Dumplings} \end{array} \begin{array}{cccc} \text{Mix (c)} & \begin{array}{c}\text{Skim}\\\text{milk (c)}\end{array} & \begin{array}{c}\text{Whole}\\\text{eggs}\end{array} & \begin{array}{c}\text{Oil}\\\text{(Tbsp)}\end{array} \\ \left[\begin{array}{cccc} 2 & 1 & 2 & 0 \\ 2 & 1\frac{1}{3} & 1 & 2 \\ 2 & \frac{3}{4} & 0 & 0 \\ 1\frac{2}{3} & \frac{2}{3} & 0 & 0 \end{array}\right] \end{array}$$

Matrix F below gives the grams of fat in each of the ingredients shown in matrix I.

$$\text{matrix } F = \begin{array}{c} \\ \text{1 c mix} \\ \text{1 c skim milk} \\ \text{1 whole egg} \\ \text{1 Tbsp oil} \end{array} \begin{array}{c} \text{Fat (g)} \\ \left[\begin{array}{c} 7.5 \\ 0.3 \\ 5.8 \\ 14.0 \end{array}\right] \end{array}$$

a. Find IF. **b.** What does matrix IF represent?

■ Solution ■

a. $IF \approx \begin{bmatrix} 2 & 1 & 2 & 0 \\ 2 & 1.33 & 1 & 2 \\ 2 & 0.75 & 0 & 0 \\ 1.67 & 0.67 & 0 & 0 \end{bmatrix} \begin{bmatrix} 7.5 \\ 0.3 \\ 5.8 \\ 14.0 \end{bmatrix}$

$= \begin{bmatrix} 2(7.5) + 1(0.3) + 2(5.8) + 0(14.0) \\ 2(7.5) + 1.33(0.3) + 1(5.8) + 2(14.0) \\ 2(7.5) + 0.75(0.3) + 0(5.8) + 0(14.0) \\ 1.67(7.5) + 0.67(0.3) + 0(5.8) + 0(14.0) \end{bmatrix}$ ← This entry is the sum of the products of the elements in the first row of I and the corresponding elements in the first column of F.

$= \begin{bmatrix} 26.9 \\ 49.2 \\ 15.2 \\ 12.7 \end{bmatrix}$

Notice that the answer is a 4×1 matrix.

(Solution continues on next page.)

■ Solution ■ *(continued)*

b. As shown below, the row labels of matrix *IF* are the same as the row labels of matrix *I* and the column labels of matrix *IF* are the same as the column labels of matrix *F*. So, matrix *IF* represents the total fat content in each of the four recipes on the baking mix box.

$$
IF = \begin{array}{c} \\ \text{Pancakes} \\ \text{Waffles} \\ \text{Biscuits} \\ \text{Dumplings} \end{array} \overset{\text{Fat (g)}}{\begin{bmatrix} 26.9 \\ 49.2 \\ 15.2 \\ 12.7 \end{bmatrix}}
$$

1. Whole milk has 8 g of fat per cup. If whole milk is substituted for equal amounts of skim milk in the recipes discussed in Example 1, which entries in matrices *I*, *F*, and *IF* would change? What would be the new entries in the affected matrices?

The carbohydrate and protein contents of the ingredients in the recipes discussed in Example 1 are given in the following table.

	Protein (g)	Carbohydrate (g)
1 c baking mix	9	84
1 c skim milk	8.9	12.6
1 whole egg	6.5	0.5
1 Tbsp oil	0	0

2. Write a matrix *N* that models the protein, carbohydrate, and fat contents of each of the ingredients. What are the dimensions of *N*?

3. How would you use matrix multiplication to find a matrix modeling the protein, carbohydrate, and fat contents of each of the recipes?

4. **Technology** Use a graphing calculator or computer software to find the matrix product described in Exercise 3. Label the rows and columns of the resulting matrix.

5. Suppose each of the recipes in matrix *I* makes four servings. How could you use scalar multiplication to find a matrix that gives the protein, carbohydrate, and fat contents per serving of each of the four recipes?

6. **Technology** Use a graphing calculator or computer software to find the matrix product described in Exercise 5. Label the rows and columns of the resulting matrix.

For Exercises 7–12, use the matrices A, B, C, D, and E to find each product. If the matrices cannot be multiplied, state that the product is _undefined_.

$$A = \begin{bmatrix} -1 \\ 3 \\ 5 \end{bmatrix} \quad B = \begin{bmatrix} 2 & 0 \\ 1 & 4 \end{bmatrix} \quad C = \begin{bmatrix} 3 & 1 & 0 \\ -1 & 3 & 0 \\ -2 & 0 & 6 \end{bmatrix} \quad D = \begin{bmatrix} 2 & 1 & 0 \\ 4 & -3 & -1 \end{bmatrix} \quad E = \begin{bmatrix} 6 & -4 \\ 0 & -3 \end{bmatrix}$$

7. AD

8. DA

9. BE

10. EB

11. $(2C) \cdot A$

12. $2 \cdot (CA)$

13. Mathematics Journal Explain how to multiply a 2×2 matrix and a 2×3 matrix. Use an example in your explanation.

· · · · · · · · · · · · · · · · · · · ·
Spiral Review

For Exercises 14–17, use matrices A and B. *(Section 1.3)*

$$A = \begin{bmatrix} 2 & 0 \\ 1 & -1 \end{bmatrix} \qquad B = \begin{bmatrix} 6 & 1 \\ 0 & 5 \end{bmatrix}$$

14. Find $A + B$.

15. Find $1.2B$.

16. Find $B - A$.

17. Find $6A - 2B$.

Express each of the following as a decimal between 0 and 1. *(Toolbox, page 788)*

18. a 15% chance

19. a 3 in 5 chance

20. a 78% chance

Solve each inequality. *(Toolbox, page 787)*

21. $3x - 7 \geq 11$

22. $4 - x > 20$

23. $\frac{1}{5}x \leq 3$

24. $-(2x + 3) < 15$

Section 1.5

Using Simulations as Models

Learn how to . . .

- model a situation with a simulation

So you can . . .

- make predictions

Application

Pilots and astronauts use flight simulators to practice maneuvers and to learn how to handle emergency situations.

Terms to Know

Example / Illustration

Simulation (p. 29)
    an experiment used to model a situation and
    make predictions

Free-throws shot by a basketball player who has a 75% success rate can be modeled using the spinner shown below. There is a 75% chance the pointer will land on "MADE" rather than "MISSED."

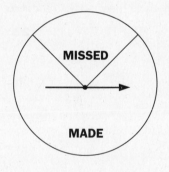

UNDERSTANDING THE MAIN IDEAS

In a simulation, you use a device, such as a spinner or a random number generator, to model an actual event about which you want to make predictions. If the simulation is done a large number of times, the probability given by the model should be about the same as the probability of the actual event.

A gardener is hoping that there will be at least one day with no rain during the coming weekend. The weather report predicts a 33% chance of rain on Saturday and a 50% chance of rain on Sunday. Describe a simulation that can be used to predict the number of days it will not rain during this coming weekend.

◼ Solution ◼

First, find a model for the event "It does not rain on Saturday."

When rolling a die, let rolling a 1 or 2 mean "It rains Saturday" and let rolling a 3, 4, 5 or 6 mean "It does not rain Saturday." Since the chance of rolling 1 or 2 is $\frac{2}{6} \approx 33\%$, and the chance of rolling any of the other numbers is $\frac{4}{6} \approx 67\%$, this is a good model for the event.

Next, find a model for the event "It does not rain Sunday."

When tossing a coin, let landing heads mean "It rains Sunday" and let landing tails mean "It does not rain Sunday." Since landing heads and landing tails each has a 50% chance of happening when tossing a coin, this is a good model for the event.

To perform the experiment, roll one die and toss one coin. For example, if you get 3, 4, 5, or 6 on the die *and* tails on the coin, this result models the event that it does not rain either day during the coming weekend. Perform the experiment numerous times, recording each result. (*Note*: The more times you repeat the experiment, the more accurate the resulting prediction wil be.)

The table below shows the results of 20 trials of the simulation described in the Example. Use this information for Exercises 1–7.

Trial number	Die result	Coin result	Number of days without rain predicted	Trial number	Die result	Coin result	Number of days without rain predicted
1	5	H	1	11	4	T	?
2	3	T	2	12	4	H	?
3	1	H	0	13	3	H	?
4	5	H	1	14	2	H	?
5	3	H	1	15	6	T	?
6	6	T	?	16	4	T	?
7	2	T	?	17	5	T	?
8	5	T	?	18	4	T	?
9	1	T	?	19	1	T	?
10	1	H	?	20	4	H	?

1. Complete the column "Number of days without rain predicted."

2. Find the mean number of days without rain predicted by adding the numbers in the last column and dividing by 20.

3. Perform 20 additional trials of the simulation described in the Example. Make a table of your results like the one shown above.

4. What is the mean number of days without rain predicted by your 20 trials?

5. What is the mean number of days without rain predicted if you combine the 20 trial results given in the table above with the results of your 20 trials recorded in Exercise 3?

6. **Writing** If you were the gardener, would you be optimistic or pessimistic about the chances of having a day without rain during this coming weekend? Explain your answer.

7. **Open-ended** Describe another method you could use to simulate the chances for no rain on Saturday and Sunday of the coming weekend.

Describe how you would simulate each situation.

8. A prize is placed in 10% of the boxes of a certain brand of cereal. You want to know, on average, how many boxes of cereal you would have to buy in order to receive a prize.

9. You are going to take a multiple-choice test containing ten questions. Each question has three choices, with only one of the three choices being correct. You wonder, on average, how many questions you would get correct by guessing the answers without looking at the questions.

10. A basketball player has a 75% success rate on free-throws. You wonder how many free-throws attempts the basketball player would need in order to make 4 free-throws in a row.

TECHNOLOGY Use a graphing calculator and the command "Int(4*rand)+1" to simulate the outcome of a free-throw shot by the basketball player described in Exercise 10. Let "1," "2," and "3" represent a made free-throw and let "4" represent a missed free-throw.

11. Repeatedly press the ENTER key on the calculator to simulate shooting free-throws, until you get four successive results that are not a "4." How many times did you have to press the ENTER key to achieve this result?

12. Repeat the experiment in Exercise 11 nine more times. Each time record the number of "attempts" it takes to get four "made free-throws" in a row. Using your 10 trials, what is the average number of attempts necessary for this basketball player to make four free-throws in a row?

Multiply. *(Section 1.4)*

13. $\begin{bmatrix} 6 & 1 \\ 3 & 4 \end{bmatrix} \begin{bmatrix} 2 \\ -2 \end{bmatrix}$

14. $\begin{bmatrix} 1 & 0 & 1 \\ 0 & 2 & 0 \\ -1 & 0 & -1 \end{bmatrix} \begin{bmatrix} 6 & 2 \\ -9 & 4 \\ 3 & -1 \end{bmatrix}$

15. $\begin{bmatrix} 2 & -3 \\ 0 & -1 \\ -4 & 0 \end{bmatrix} \begin{bmatrix} 1 & 6 & 3 \\ -1 & 8 & 0 \end{bmatrix}$

For each equation, make a table of values for *x* and *y*. Use –3, –2, –1, 0, 1, 2, and 3 as values for *x*. Then use your points to graph each equation.
(Toolbox, page 795)

16. $y = 6x$

17. $y = -2x$

18. $y = 5x + 7$

19. $y = -3x - 8$

Chapter 1 Review ....

Complete these exercises for a review of Chapter 1. If you have difficulty with a particular problem, review the indicated section.

The table shows the average number of households that watched the top-rated television series in the United States from 1985 to 1990.

Year season began (September–April)	Top-rated series	Number of households watching (millions)
1985	Bill Cosby Show	85.9
1986	Bill Cosby Show	87.4
1987	Bill Cosby Show	88.6
1988	Roseanne	90.4
1989	Roseanne	92.1
1990	Cheers	93.1

1. What was the average change in the number of viewing households in each year? *(Section 1.1)*

2. What was the average percent change in the number of viewing households each year? *(Section 1.1)*

3. Use a constant growth model and a proportional growth model to predict the number of households watching the top-rated series during the 1991 television season. *(Section 1.1)*

4. Write a linear function modeling the number of households watching the top-rated series since 1985. Then write an exponential function modeling the situation. *(Section 1.2)*

5. Use the models from Exercise 4 to predict the number of households watching the top-rated series in the year 2000. *(Section 1.2)*

6. **Writing** What do you think could account for the growth shown by the data in the table? *(Section 1.2)*

Use the matrices below to evaluate each matrix expression. If an operation cannot be performed, write *undefined*. *(Sections 1.3 and 1.4)*

$$A = \begin{bmatrix} 4 & 3 & 0 \\ 0 & -1 & 5 \end{bmatrix} \quad B = \begin{bmatrix} -1 & 7 \\ 5 & -2 \\ 2 & 8 \end{bmatrix} \quad C = \begin{bmatrix} 5 & 1 & -8 \\ -5 & 6 & 0 \end{bmatrix} \quad D = \begin{bmatrix} 14 & 0 & 1 & 6 & 8 \\ 2 & -1 & 0 & 0 & 7 \\ 5 & 1 & 0 & 8 & 4 \end{bmatrix}$$

7. $2A - C$ **8.** $A + B$ **9.** DC **10.** CD **11.** $-BC$

12. **Open-ended** Give four examples of possible dimensions for matrices R and S so that the product RS is a 3×2 matrix. *(Section 1.4)*

Mike's Diner is having a special promotion. Every time you eat there, you receive a scratch card with one of the letters M, I, K, or E printed on it. If you collect all four letters, you win a free meal. The cards are printed so that 30% have M on them, 20% have I on them, 40% have K on them, and 10% have E on them.

13. Describe how you would use random numbers to simulate this special promotion. *(Section 1.5)*

14. The table below shows the results of an actual simulation to determine how many cards you would need to collect in order to win a free meal at Mike's Diner. Based on these results, how many cards (on average) would you need in order to win? *(Section 1.5)*

Simulation	Results
1	M, I, E, K
2	I, M, E, K
3	K, M, K, K, I, K, K, I, K, K, E
4	K, M, E, I
5	M, I, M, I, K, K, I, I, I, M, K, K, M, K, I, I, K, M, I, I, K, K, M, I, K, M, K, M, I, I, K, M, K, E
6	K, M, I, K, M, M, M, I, K, I, K, K, I, M, I, M, K, K, I, K, K, M, K, I, I, I, K, K, I, K, K, I, I, K, K, M, K, K, M, K, I, K, K, M, K, K, I, K, M, K, K, M, I, M, M, I, K, K, I, K, I, E

1. Make a histogram of the television series data given for Exercises 1–6 in the Chapter Check-Up. Include a bar for the predicted 1991 data.

2. Find the mean, the median, and the mode of the number of households watching the top-rated series each year from 1985 to 1990.

Solve each equation.

3. $4(x - 7) = -24$

4. $\dfrac{q + 8}{4} = 1.3$

5. $\dfrac{2 - 5x}{x + 2} = 7$

6. $2y + 8 = 9y + 4$

Evaluate each expression when $a = 3$ and $b = -5$.

7. $(a - b)^2$

8. $(a - b)(a + b)$

9. $a^2 + b^2$

Direct Variation

Application

The price you pay for gasoline varies directly with the amount you buy. The price per gallon is the constant of variation.

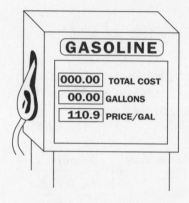

Learn how to . . .

- recognize direct variation
- write and use direct variation equations

So you can . . .

- analyze data and make predictions

Terms to Know

Example / Illustration

Direct variation (p. 46)	
a situation in which the ratio of two variables is constant	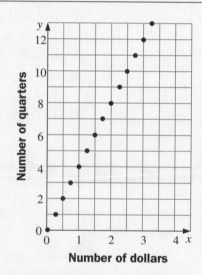

The ratio of quarters to dollar bills for equivalent amounts of money is 4 to 1. If x = the number of dollar bills and y = the number of quarters, this ratio can be written as $\frac{y}{x} = 4$, or $y = 4x$.

Constant of variation (p. 46)
the constant ratio of two variables in a direct variation (In the direct variation equation $y = ax$, $a \neq 0$, a is the constant of variation.)

The constant of variation in the ratio above is 4.

Slope (p. 47) for any line, the ratio $\dfrac{\text{vertical change}}{\text{horizontal change}}$ (In a direct variation, the slope is the constant of variation.)	The slope of the graph of the direct variation equation $y = -2x$ is -2.

UNDERSTANDING THE MAIN IDEAS

Direct variation

Direct variation can be modeled by an equation of the form $y = ax$. The graph of a direct variation equation is a line with slope a that passes through the origin.

Example 1

For each table, decide if the variables are in direct variation. If so, write an equation relating the two variables.

a. Dimensions of balls

Type of ball	Circumference	Surface area
Golf	5 in.	8 in.^{2}
Baseball	9 in.	26 in.^{2}
Soccer	28 in.	250 in.^{2}
Basketball	30 in.	286 in.^{2}

b. Vacation mileage

Day	Time	Distance
Saturday	6 h	310 mi
Sunday	4 h	209 mi
Monday	7.5 h	392 mi
Friday	5.5 h	285 mi

■ Solution ■

a. Find the ratio of *surface area to circumference* for each type of ball.

Type of ball	Surface area / Circumference
Golf	$\dfrac{8}{5} = 1.6$
Baseball	$\dfrac{26}{9} \approx 2.89$
Soccer	$\dfrac{250}{28} \approx 8.93$
Basketball	$\dfrac{286}{30} \approx 9.53$

All the ratios are different, so there is no common ratio. Therefore, the surface area and circumference of the balls are not in direct variation.

b. Find the ratio of *distance to time* for each day.

Day	Distance / Time
Saturday	$\dfrac{310}{6} \approx 51.67$
Sunday	$\dfrac{209}{4} \approx 52.25$
Monday	$\dfrac{392}{7.5} \approx 52.27$
Friday	$\dfrac{285}{5.5} \approx 51.82$

The ratios, while not exactly the same, are all very close to 52. So, distance and time do have a common ratio of about 52. A direct variation equation relating distance and time is

$$\frac{D}{t} = 52 \text{ or } D = 52t.$$

---

1. In part (b) of the Example, what does the common ratio represent?

2. Make a scatter plot of the data given in part (a) of the Example. How can you tell from the scatter plot that the two variables are not in direct variation?

3. Make a scatter plot of the data given in part (b) of the Example. On the same set of axes, graph the direct variation equation $D = 52t$. Is the line a good fit for the data?

For Exercises 4–6, refer to the table at the right.

4. Do diameter and circumference have a common ratio? If so, write a direct variation equation relating diameter and circumference.

5. **Writing** A tennis ball has a circumference of 8 in. and a diameter of 2.55 in. Does a tennis ball fit the relationship shown in the table? Explain your answer.

6. The circumference of a table tennis ball is 4.5 in. Predict the diameter of a table tennis ball.

7. The diameter of a softball is 3.8 in. What is its circumference?

Dimensions of Balls		
Type of ball	**Circumference**	**Diameter**
Golf	5 in.	1.59 in.
Baseball	9 in.	2.86 in.
Soccer	28 in.	8.91 in.
Basketball	30 in.	9.55 in.

For each graph, tell whether y varies directly with x. If so, give an equation for the graph.

8.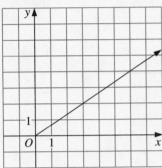

9.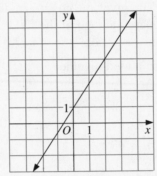

10.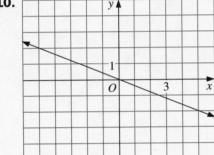

11.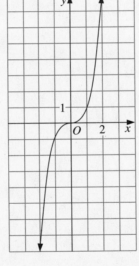

For each equation, tell whether y varies directly with x. If so, graph the equation.

12. $y = \dfrac{x}{9}$

13. $y = \dfrac{5}{x}$

14. $y = 6x - 3$

15. $\dfrac{y}{x} = 2.5$

16. The amount of water in a graduated cylinder varies directly with the height of the water. When the depth of the water is 5 cm, there is 150 ml of water in the cylinder. Write a direct variation equation relating the height of the water (in cm) to the volume of water (in ml).

17. Graph the equation you wrote for Exercise 16. Mark the point on the graph that shows the height of the water when the volume is 225 ml.

·····················
Spiral Review

18. For 25 cents, a gumball machine dispenses one of 3 different colors of giant gumballs. Each of the colors is equally likely to drop from the machine. On average, how much would you have to spend to get gumballs of all three different colors? Use a simulation to answer this question. *(Section 1.5)*

19. The linear function $C = 100 + 1.6y$ gives the cost C, in dollars, for reseeding a grassy field in terms of the square yardage of the field y. Tell what the numbers 100 and 1.6 mean in this situation. *(Section 1.2)*

20. Which pair(s) of matrices can be added together? Which pair(s) can be multiplied? *(Sections 1.3, 1.4)*

$$A = \begin{bmatrix} 4 & 2 \\ -3 & 8 \\ 0 & 5 \end{bmatrix} \qquad B = \begin{bmatrix} 6 & -1 & 2 \\ 3 & 9 & 4 \end{bmatrix} \qquad C = \begin{bmatrix} 2 & 0 \\ 9 & 5 \\ 6 & -4 \end{bmatrix}$$

Study Guide, ALGEBRA 2: EXPLORATIONS AND APPLICATIONS

Linear Equations and Slope-Intercept Form

Learn how to . . .

- write an equation of a line in slope-intercept form
- graph a linear equation in slope-intercept form

So you can . . .

- model events

Application

The length of a nail is described in "pennies," abbreviated "d." For nail sizes from 2d (read "two penny") up to 10d, the relationship between "pennies" and inches is modeled by a linear equation with slope 0.25 and intercept 0.5 as shown in the figure.

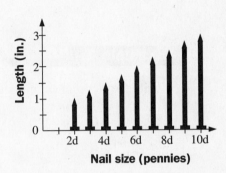

Terms to Know

Example / Illustration

Vertical intercept (p. 53) the place where a graph crosses the vertical axis	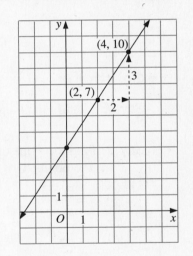 The vertical intercept of this graph is 4.
Slope-intercept form (p. 53) the form of an equation of a line, $y = ax + b$, where a is the slope of the line and b is its vertical intercept	The slope-intercept form of the equation of the line graphed above is $y = \frac{3}{2}x + 4$.

Relating a slope-intercept equation to its graph

Recall that to find the slope of a line, you find the ratio of the vertical change to the horizontal change. In general, if (x_1, y_1) and (x_2, y_2) are two points on a line, then the slope a of the line in the slope-intercept form of the equation is:

$$a = \frac{y_2 - y_1}{x_2 - x_1}$$

Example 1

Write the slope-intercept equation of the line.

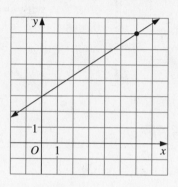

■ Solution ■

Step 1: Identify the y-intercept.

The graph crosses the y-axis at the point $(0, 3)$, so the y-intercept is 3.

Step 2: Find the slope of the line.

The points $(0, 3)$ and $(6, 7)$ lie on the line. Using the formula,

$$a = \frac{y_2 - y_1}{x_2 - x_1} \qquad \leftarrow \text{Let } (x_1, y_1) = (0, 3) \text{ and } (x_2, y_2) = (6, 7).$$

$$= \frac{7 - 3}{6 - 0} \qquad \leftarrow \text{Substitute 0 for } x_1, \text{ 3 for } y_1, \text{ 6 for } x_2, \text{ and 7 for } y_2.$$

$$= \frac{4}{6}, \text{ or } \frac{2}{3}$$

(*Note:* You can verify that your calculation of the slope is reasonable by examining the graph. If the line rises as you move from left to right along the horizontal axis, then its slope is positive; if the line falls as you move from left to right along the horizontal axis, then its slope is negative.)

Step 3: Write the equation of the line using the slope-intercept form.

$$y = ax + b \qquad \leftarrow b = 3 \text{ (from Step 1) and } a = \frac{2}{3} \text{ (from Step 2)}$$

$$y = \frac{2}{3}x + 3$$

Find the slope-intercept equation of each line.

1.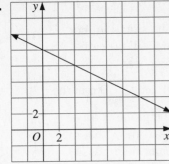

2.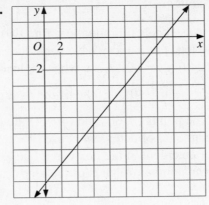

3. the line through the points (0, 6) and (–2, 3)

4. the line through the points (0, –4) and (8, 0)

State the slope and *y*-intercept of each line with the given equation.

5. $y = -1.6x + 2.4$ **6.** $y = x - 6$ **7.** $y = 10 - 4x$

Using the slope-intercept form to model situations

Many real-world situations can be modeled by linear equations in slope-intercept form.

Example 2

Model the following situation with a graph and an equation.

> A new "super size" box of laundry detergent contains 60 c of detergent. Suppose a typical load of laundry requires $\frac{3}{4}$ c of detergent. What is the relationship between the amount of laundry detergent remaining in the box and the number of loads of laundry washed?

▪ Solution ▪

The amount of laundry detergent left in the box depends on the number of loads of laundry that have been done. The independent variable, *x*, is the number of loads of laundry. The dependent variable, *y*, is the amount of detergent left in the box.

(Solution continued on next page.)

To draw a graph that models this situation, find two points that fit the situation. For example, when no loads of laundry have been done ($x = 0$), there are still 60 c of detergent in the box (so $y = 60$). Also, when 40 loads of laundry have been done ($x = 40$), there have been $\frac{3}{4}(40)$, or 30 c of the detergent used, which leaves 30 c in the box ($y = 30$). So, the points (0, 60) and (40, 30) lie on the graph. The graph is linear because the same amount of detergent is used for every load. Also, we know that the number of cups of detergent in the box cannot be less than 0 or greater than 60, so $0 \le y \le 60$, and the number of loads of laundry cannot be less than 0, so $0 \le x$.

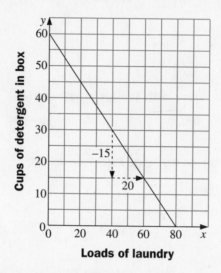

From the graph, notice that the slope is $-\frac{3}{4}$. This is reasonable because $\frac{3}{4}$ c of detergent is *used* for each load of laundry, so the amount of detergent in the box is decreasing. Also, the y-intercept is 60. Therefore, an equation (in slope-intercept form) that models this situation is

$$y = -\frac{3}{4}x + 60.$$

Model each situation with an equation and a graph. Be sure to identify the independent and dependent variables.

8. A new jar of peanut butter contains 50 tablespoons. Each peanut butter sandwich requires 2 tablespoons of peanut butter.

9. The cost of holding a party at the Funtime Arcade is $40 plus $8 per guest.

10. A new member of Books-by-Mail receives 4 books. After that, the member receives one book each month.

11. The temperature at sundown was 48°F. It is dropping 3 degrees per hour.

For Exercises 12–14, graph each equation.

12. $y = 0.8x + 3$ **13.** $y = 6 - \dfrac{3}{5}x$ **14.** $y = -2 - 2x$

15. Mathematics Journal Explain, with illustrations, why the slope of the graph of a direct variation is the same as the common ratio.

....
Spiral Review

16. Parking in the Center City garage costs $3 per hour. Write a direct variation equation that relates the number of hours a car is parked in the garage to the parking fee. *(Section 2.1)*

17. Which of the following expressions is equivalent to $7 - x$? *(Toolbox, page 781)*

    A. $5 - (x + 2)$ **B.** $5 + (x - 2)$ **C.** $5 - (x - 2)$

Solve each proportion. *(Toolbox, page 785)*

18. $\dfrac{5 + x}{5 - x} = 4$ **19.** $\dfrac{3}{2 - 4x} = \dfrac{1}{6}$ **20.** $2 = \dfrac{x + 5}{2x - 14}$

Point-Slope Form and Function Notation

Learn how to . . .

- write the point-slope equation of a line
- use function notation
- find the domain and range of a linear function

- **So you can . . .**

  make predictions

Application

If you drive about the same number of miles each week, you can use two different odometer readings to write a function that models your total miles driven over time.

Terms to Know

Example / Illustration

Point-slope equation (p. 59)

the form of the equation of a line, $y = y_1 + a(x - x_1)$, where a is the slope of the line and (x_1, y_1) is a point on the line

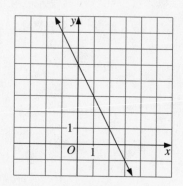

The point-slope equation of this line is $y = 1 - 2(x - 2)$.

Function notation (p. 60)

notation of the form $y = f(x)$, which tells you that y is a function of x

$y = f(x) = 10 - 2x$

$y = f(x) = -3 + 4(x + 1)$

Domain (p. 60)

the set of values of the independent variable for which a function is defined

If the domain of the function $f(x) = 10 - 2x$ is $0 \le x \le 5$, then the range is $0 \le y \le 10$.

Range (p. 60)

the set of all values of $f(x)$, where x is in the domain of f

---

Understanding the Main Ideas

When you know (or can determine) the slope of a line and one point on the line, you can write the point-slope form of the equation of the line.

Example 1

Kareem is cataloging books for the library at the rate of 25 books per day. At the end of the twelfth day, there are 475 books left to be cataloged.

a. Write and graph a linear equation modeling the number of books left to be cataloged.

b. How many days will Kareem have to work to complete the cataloging?

■ Solution ■

a. The number of books Kareem has left to catalog depends on how many days he has worked. So the independent variable, x, is the number of days he has been cataloging and the dependent variable, y, is the number of books yet to be cataloged.

Write an equation.

We know that after 12 days there are still 475 books to be cataloged, so the point (12, 475) must satisfy the equation. Since Kareem is cataloging books at a rate of 25 books per day, the number of books yet to be cataloged is *decreasing* by 25 each day, so the slope of the line must be –25. Therefore, the equation (in point-slope form) is

$$y = y_1 + a(x - x_1) \qquad \leftarrow (x_1, y_1) = (12, 475), a = -25$$
$$y = 475 - 25(x - 12)$$

Graph the equation.

First plot the point (12, 475). Now use this point and the slope, –25, to find and plot at least two more points that satisfy the equation. A slope of –25 means that for every *increase* of 1 in the value of x, there is a corresponding *decrease* of 25 in the value of y. So, the point (12 + 1, 475 – 25), or (13, 450) satisfies the equation. Likewise, the point (12 + 10, 475 – 250), or (22, 225) satisfies the equation. A line can be drawn through these three points. However, since the number of days and the number of books cannot be less than 0, the graph of the equation for this situation is restricted to the portion of the line that lies in Quadrant I. The graph is shown on the next page.

(Solution continues on next page.)

■ **Solution** ■ *(continued)*

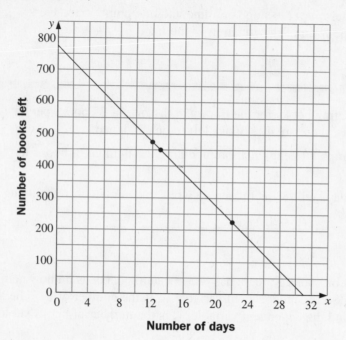

Number of days

b. The cataloging will be complete when the number of books is 0; that is, when $y = 0$. From the graph you can see that $y = 0$ when $x = 31$. This can be verified by substituting 0 for y in the equation and solving for x. So, it will take Kareem 31 days to complete the cataloging.

1. Check the answer to part (b) of Example 1 by substituting $x = 31$ and $y = 0$ into the equation of the line found in part (a).

2. How many books were there to be cataloged when Kareem began? Where is this information shown on the graph?

3. How would the point-slope equation change if Kareem had been cataloging 30 books per day instead of 25 books?

4. How would the point-slope equation change if Kareem was cataloging at the rate of 25 books per day, but after 12 days he still had 325 books left to be cataloged instead of 475 books?

State the slope and one point on each line. Then graph the equation.

5. $y = 35 + 1.5(x - 7)$

6. $y = 8 + 3(x + 12)$

For Exercises 7–9:

a. Write a point-slope equation of the line passing through the given point and having the given slope.

b. Graph the equation.

c. Write the equation in slope-intercept form.

7. point: $(-4, 6)$

slope: -1

8. point: $(1, -3)$

slope: $\dfrac{5}{6}$

9. point: $(0, 9)$

slope: $\dfrac{2}{3}$

You can also write the point-slope equation of a line when you know two points that lie on the line.

Example 2

Write a point-slope equation of the line through the points $(12, 3)$ and $(20, -2)$.

■ Solution ■

Step 1: Find the slope of the line.

$$a = \frac{y_2 - y_1}{x_2 - x_1} \quad \leftarrow \text{Let } (x_1, y_1) = (12, 3) \text{ and } (x_2, y_2) = (20, -2).$$

$$= \frac{-2 - 3}{20 - 12}$$

$$= -\frac{5}{8}$$

(*Note:* In the slope formula, you can let either point be (x_1, y_1) and the other be (x_2, y_2).)

Step 2: Use the slope and either point to write the equation of the line.

Using $(12, 3)$ as (x_1, y_1):

$$y = 3 + \left(-\frac{5}{8}\right)(x - 12)$$

$$y = 3 - \frac{5}{8}(x - 12)$$

Using $(20, -2)$ as (x_1, y_1):

$$y = -2 + \left(-\frac{5}{8}\right)(x - 20)$$

$$y = -2 - \frac{5}{8}(x - 20)$$

Each equation is a correct equation for the line.

10. Technology Use a graphing calculator to show that the two equations found in Step 2 of Example 2 represent the same line.

Write a point-slope equation of the line passing through the given points.

11. $(3, -7)$ and $(1, -1)$

12. $(12, 0)$ and $(6, 3)$

13. $(-2, 6)$ and $(5, 6)$

The domain of a function f is the set of values of x for which f is defined. The range of a function f is the set of all values of $f(x)$, where x is in the domain of f.

Example 3

The function $f(x) = -\frac{3}{5}x + 6$ is defined for $x \geq -5$.

   a. Find $f(20)$.   **b.** Find the domain and range of f.

■ Solution ■

a. $f(x) = -\frac{3}{5}x + 6$ ← Substitute 20 for x.

$= -\frac{3}{5}(20) + 6$

$= -12 + 6$

$= -6$

b. The domain is given as $x \geq -5$. Now use the domain to find the range by transforming the left side of the inequality $x \geq -5$ into $-\frac{3}{5}x + 6$.

$x \geq -5$ ← Multiplying both sides of the inequality by $-\frac{3}{5}$.

$-\frac{3}{5}x \leq -\frac{3}{5}(-5)$ ←   Multiplying by a negative number changes the direction of the inequality symbol.

$-\frac{3}{5}x \leq 3$ ← Add 6 to both sides of the inequality.

$-\frac{3}{5}x + 6 \leq 3 + 6$

$-\frac{3}{5}x + 6 \leq 9$

So $f(x) \leq 9$. This is the range of the function.

Find the domain and range of each function.

14. $f(x) = \frac{2}{3}x + \frac{26}{3}$ for $x \geq 2$ **15.** $f(x) = \frac{4}{5}x - 2$ for $x \geq 0$

Danika bought a used car. After 10 weeks of driving, the odometer read 29265. After 25 weeks, it read 32865. Danika knows she drives about the same number of miles each week.

16. Write a point-slope equation that models r, the odometer reading, as a function of t, the number of weeks she has owned the car.

17. What was the odometer reading when Danika bought the car?

Spiral Review

For Exercises 18–20, find the slope-intercept equation of the line passing through each pair of points. *(Section 2.2)*

18. (0, 5) and (2, 3)   **19.** (0, –2) and (4, 2)   **20.** (5, 0) and (0, –4)

21. A school nurse is keeping weekly records on the number of new cases of influenza in the school district. The data appear in the table. Write a linear equation to model the number of new cases of influenza each week. *(Section 1.2)*

Week	Number of cases
0	5
1	16
2	28
3	39

Solve each equation. *(Toolbox, p. 784)*

22. $x - \dfrac{2}{5} = \dfrac{9}{10}$

23. $15 - x = 2x + 8$

24. $16n = 128$

25. $\dfrac{r}{9} = 3r + 52$

Fitting Lines to Data

Learn how to . . .

- draw a line of
  fit and find
  its equation

So you can . . .

- make predictions

Application

The equation of a line of fit
provides a linear growth
model for a set of
data points.

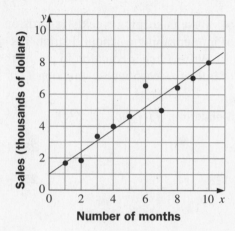

Terms to Know

Least-squares line (p. 67)
    the standard line of fit for a set of data points

Example / Illustration

A graphics calculator can be
used to find the equation of the
least-squares regression line
(often called just the
least-squares line) for a set of
data points. Most graphing
calculators use the term *LinReg*
(for "linear regression") for this
statistical function.

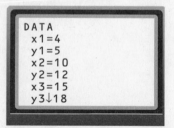

The equation of the least-squares
line given for these data points is
$y = 1.1x + 0.8$.

UNDERSTANDING THE MAIN IDEAS

When fitting a line to a group of points on a scatter plot, you choose a position
for the line that best represents all the points.

Example

The table at the right gives the number of active
pilot's certificates held in the United States in
various years.

a. Make a scatter plot of the data.

b. If the data have a linear relationship, draw
a line of fit on the scatter plot and determine
an equation for the line.

Year	Number of pilot's certificates
1985	720,000
1990	700,000
1991	690,000
1992	680,000
1993	670,000

Solution

a. Let t be the number of years since 1980 and let y be the number of pilot's
certificates in thousands. Plot the points (t, y) on a coordinate plane.

t	y
5	720
10	700
11	690
12	680
13	670

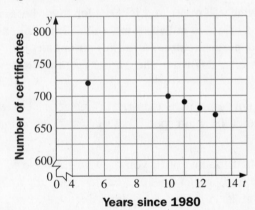

Years since 1980

b. The data points do appear
to have a linear
relationship. A line of fit
has been added to the
scatter plot, as shown at
the right. Since not every
person would draw exactly
this same line of fit for
these data points, this line
is just one possible fitted
line. However, every
reasonable fitted line
would have a negative slope.

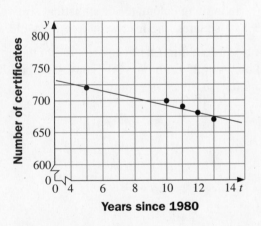

Years since 1980

(Solution continues on next page.)

■ Solution ■ *(continued)*

To write an equation for this line of fit, first find the coordinates of two points that appear to lie on the line, for example, (5, 720) and (12, 680). Using these two points, the slope of the line is

$$a = \frac{680 - 720}{12 - 5} = \frac{-40}{7} \approx -5.7$$

So, a point-slope equation of the line of fit, using the point (5, 720), is
$y = 720 - 5.7(x - 5)$.

1. Use the equation in the Example to predict the number of pilot's certificates in the year 2000.

2. **Open-ended** Find a point-slope equation for another line of fit that seems to reasonably fit the data in the scatter plot.

3. **Technology** Use a graphing calculator or statistical software to find the least-squares line for the data in the Example.

4. **Technology** Graph the line in the Example on the same set of axes as the line of fit found in Exercise 3. How do they compare?

On a cold day, the wind will make you feel colder than the actual air temperature. This effect is called "wind chill." The table at the right gives the apparent temperature (in Fahrenheit) for various wind speeds on a day when the actual air temperature is 30°.

Wind speed (mi/h)	Apparent temperature (°F)
10	16
15	9
20	4
30	-2
40	-5

5. Make a scatter plot of the data and draw a line of fit. Use the line of fit to predict the wind speed that makes the apparent temperature 0°F.

6. Find an equation of the line of fit you drew in Exercise 5. Use your equation to predict the apparent temperature when the wind blows at 25 mi/h on a 30°F day.

7. **Technology** Use a graphing calculator or statistical software to find an equation of the least-squares line for the data in the table.

8. Use the equation you found in Exercise 7 to predict the apparent temperature when the wind blows at 25 mi/h on a 30°F day. Compare your result with your answer to Exercise 6.

The table at the right shows the light output, in lumens, of new frosted incandescent light bulbs of various wattages.

Wattage	Light output (lumens)
15	125
25	190
40	505
60	1000
75	1190
100	1750
150	2850

9. Make a scatter plot of the data and draw a line of fit.

10. Find an equation of the line of fit you drew in Exercise 9.

11. Use your equation from Exercise 10 to predict the light output of a 300-watt bulb.

12. **Writing** The light output of a new 300-watt bulb is 6360 lumens. How does this compare to your prediction in Exercise 11? Explain any difference.

Spiral Review

13. The table of men's hat sizes gives the circumference, in inches, of the sweatband of a hat of each size. Make a scatter plot of the data. Then tell whether it seems to make sense to fit a line to the data. *(Toolbox, page 795)*

Hat size	Circumference (in.)
6	19
$6\frac{1}{4}$	$19\frac{3}{4}$
7	22
$7\frac{3}{4}$	$24\frac{1}{4}$
8	25

Write a point-slope equation of the line passing through the given point and having the given slope. *(Section 2.3)*

14. point: (2, 5)
    slope: 1.4

15. point: (−5, 0)
    slope: −2

16. point: (13, −4)
    slope: 1.75

For each equation, state whether y varies directly with x. *(Section 2.1)*

17. $y = \dfrac{x}{7}$

18. $y = 5 + x$

19. $y = -6x$

20. $y = 4(x - 12)$

The Correlation Coefficient

Learn how to . . .

- interpret a
  correlation
  coefficient

- recognize positive
  and negative
  correlation

So you can . . .

- determine the
  strength of a linear
  relationship

Application

The correlation coefficient tells you if a linear model provides a good description of a set of data.

Terms to Know

Example / Illustration

Correlation coefficient (p. 72) a number r between -1 and $+1$ that measures how well data points line up	$-1 \le r \le 1$
Positive correlation (p. 72) the correlation between the variables x and y when y tends to increase as x increases (For a positive correlation, the value of r is $0 < r \le 1$.)	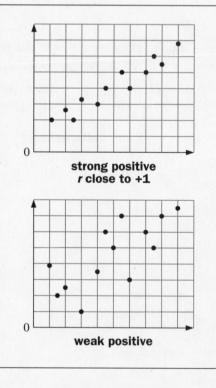 **strong positive** **r close to +1** **weak positive**

Negative correlation (p. 72)

the correlation between the variables x and y when y tends to decrease as x increases (For a negative correlation, the value of r is $-1 \leq r < 0$.)

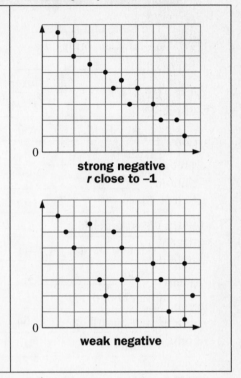

strong negative
r close to –1

weak negative

UNDERSTANDING THE MAIN IDEAS

Correlation coefficient

The correlation coefficient can tell you the strength of the *linear* relationship between two variables. A correlation coefficient close to +1 or –1 indicates a strong correlation between the variables. A correlation coefficient near 0 indicates a weak correlation between the variables. However, the correlation coefficient *does not* tell you if a change in one variable *causes* a change in the other. Also, it *does not* tell you if there is a better nonlinear relationship between the variables.

The table and scatter plot below give the average January high and low temperatures for ten California cities.

City	January Temps. (°F)	
	High	Low
Bakersfield	57	37
Bishop	53	21
Chico	54	36
Fresno	55	36
Los Angeles	65	47
Mount Shasta	41	25
Redding	54	37
San Diego	65	46
Santa Barbara	64	43
Willits	49	31

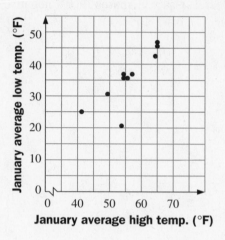

a. Use a graphing calculator or statistical software to find the correlation coefficient for the data shown in the scatter plot.

b. What does the correlation coefficient tell you about the relationship between the high and low temperatures for these cities?

c. Explain what might account for this correlation value.

Solution

a. Enter the temperature data pairs (high, low) into a graphing calculator or statistical software. Then select the linear regression (LinReg) function. The resulting calculator screen shows that the correlation coefficient r is about 0.85. (*Note*: The equation of the least-squares line is $y = ax + b$, for the values of a and b given in the calculator screen.)

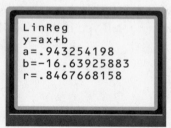

b. Since the correlation coefficient, 0.85, is close to 1, there is a strong positive correlation between the high and low temperatures in January for these cities.

c. High temperatures do not cause low temperatures, and low temperatures do not cause high temperatures. Rather, the things that do affect the January weather in these cities (altitude, longitude, proximity to the ocean, and so on), affect both the high and the low temperatures.

The table and scatter plot below give the January high temperature and the
July high temperature for the same ten California cities listed in the table in
the Example.

City	High Temps. (°F)	
	January	July
Bakersfield	57	100
Bishop	53	97
Chico	54	98
Fresno	55	99
Los Angeles	65	83
Mount Shasta	41	85
Redding	54	97
San Diego	65	76
Santa Barbara	64	76
Willits	49	86

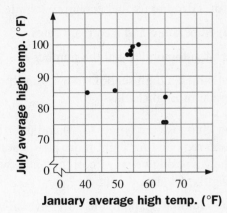

1. Do you think there is a strong linear correlation between the January high
   temperatures and the July high temperatures? Estimate the correlation
   coefficient. Will it be positive or negative?

2. **Technology** Use a graphing calculator or computer software to find the
   correlation coefficient. Is the correlation *strong* or *weak*? Is it *positive* or
   negative?

3. **Writing** Explain what might account for the correlation in Exercise 2.

Match each scatter plot to one of these values of *r* : –0.05, –0.3, 0.9, 0.4, –1.

4.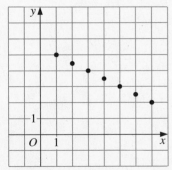

5.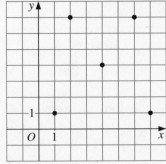

6.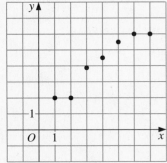

7.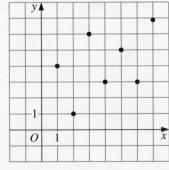

For Exercises 8–11, tell whether you would expect the correlation between the two quantities to be *positive*, *negative*, or *about zero*.

8. the weight of a car and its fuel economy in miles per gallon

9. the outside temperature and ice cream sales

10. the number of automobiles and the number of school children in a city

11. the speed at which a high school student runs a quarter mile and the student's grade in English class

12. **Open-ended** Suppose the points (2, 8) and (5, 10) both lie on a scatter plot. Write the ordered pairs for eight more points that when added to the scatter plot will show a strong negative correlation between x and y. Draw a scatter plot showing the ten data points.

13. **Mathematics Journal** Explain the difference in meaning between *correlation* and *causation*. Give examples to support your explanation.

....
Spiral Review

The table shows the average weight, in pounds, of three foods consumed by Americans in each of three years. For example, the average American consumed 30 lb of eggs in 1992. (*Section 2.4*)

Year	Consumption (pounds per person)		
	Eggs	**Milk and cream**	**Fruit**
1991	30	233	290
1992	30	231	292
1993	30	227	309

14. For each food, make a scatter plot of the consumption over the three years. Then draw a line of fit. (*Hint*: Label each horizontal axis as "Years after 1990.")

15. Find an equation for each line you drew in Exercise 14.

16. Predict the average American's consumption of each of the three foods in 1995 and 2000.

Write an equation of the line with the given slope and y-intercept.
(*Section 2.2*)

17. slope = $\frac{3}{8}$
    y-intercept = -3

18. slope = -1
    y-intercept = 12

19. slope = 0
    y-intercept = 5

Solve each linear equation. (*Toolbox, page 784*)

20. $4 + 6r = -14$

21. $\frac{x-7}{3} = 5$

22. $4(5 + s) = 100$

Linear Parametric Equations

Learn how to . . .

- write and graph a pair of linear parametic equations

- rewrite a pair of parametric equations as a single equation

So you can . . .

- model situations

Application

Physicists model the path of the baseball using parametric equations, one for the horizontal force of the throw and one for the vertical (downward) force of gravity.

Terms to Know	Example / Illustration
Parametric equations (p. 79) a pair of equations expressing the relationship between two variables in terms of a third, independent variable	$x = 2 + t$ $y = \frac{1}{2}t, \ t \geq 0$
Parameter (p. 79) the third, independent variable in a pair of parametric equations	In the parametric equations above, t is the parameter.

UNDERSTANDING THE MAIN IDEAS

Parametric equations are used to describe the position of an object in terms of time, as shown in the following example.

Example

You fly over your house in an airplane headed due north at 240 mi/h. The wind at the airplane's altitude is blowing directly from the east at 15 mi/h.

a. Using your house as the origin, write and graph parametric equations describing your path.

b. In relationship to your house, where will your airplane be 40 min after passing over the house?

■ Solution ■

a. Let x = the airplane's distance east of your house and let y = the airplane's distance north of your house. Write two equations, one expressing x as a function of t and the other expressing y as a function of t.

The airplane is flying due north at a speed of 240 mi/h, so $y = 240t$, where t is the amount of time in hours after it passed over your house.

Since the wind is blowing from the east at 15mi/h, it is pushing the airplane *west* at 15 mi/h, so $x = -15t$. (*Note:* Since x = the airplane's distance east of your house, you must use negative values for x to indicate distances to the west.)

So the parametric equations are $x = -15t$ and $y = 240t$ for $t \geq 0$ (where $t = 0$ is the time at which the airplane is directly over your house).

Now use a table of values to plot a few points (x, y). The graph is shown at the right below.

t	$x = -15t$	$y = 240t$
0	0	0
0.5	-7.5	120
1	-15	240
1.5	-22.5	360
2	-30	480

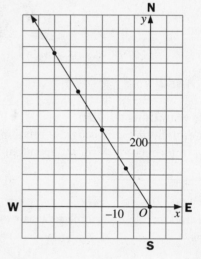

b. To find the position of the airplane 40 min (or $\frac{2}{3}$ h) after it passed over the house, substitute $\frac{2}{3}$ for t in the equations $x = -15t$ and $y = 240t$.

$$x = -15t\left(\frac{2}{3}\right) = -10 \qquad y = 240\left(\frac{2}{3}\right) = 160$$

So, after 40 min. the airplane is 10 mi west and 160 mi north of your house.

For Exercises 1–4, refer to the situation described in the Example.

1. How far north will the airplane be when it is 20 mi west of your house?

2. How far west will the airplane be when it is 640 mi north of your house?

3. Lookout Mountain is 400 mi north and 25 mi west of your house. Will the airplane pass over it?

4. What value should you use for t to find the airplane's position 20 min *before* you flew over your house? Find the airplane's position at this time.

5. Using the parametric equations found in part (a) of the Example, write an expression for y in terms of x. You must include the restriction on the value of x created by the given restriction on t. (*Hint:* Use substitution.)

A football field (including the end zones) is 360 ft long and 160 ft wide. During the half-time show of a football game, the drum major starts at one corner of the field and marches at a steady pace diagonally to the opposite corner, arriving there 2 min. later.

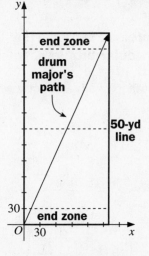

6. Use a coordinate system with the drum major's starting point in the corner of an end zone as the origin, as shown in the figure at the right. Write a pair of parametric equations that describe the path of the drum major across the field in terms of *t*, the number of minutes after the drum major begins marching.

7. At the same time the drum major begins marching, a trumpet player on the sideline starts marching directly along the 50-yd line at the rate of 75 ft/min. Write parametric equations describing the path of the trumpet player.

8. Will the drum major and the trumpet player collide? Explain your reasoning.

9. Write an equation for the path of the drum major as a function of *y* in terms of *x*. Write an equation for the path of the trumpet player as a function of *y* in terms of *x*. Are there any restrictions on the value of *x* for either equation? If so, state the restrictions.

For Exercises 10–12, graph each pair of parametric equations for the given restrictions on *t*.

10. $x = 4t$
    $y = 5t$
    $t \geq 0$

11. $x = t + 2$
    $y = t - 2$
    $5 \leq t \leq 10$

12. $x = 0.25t$
    $y = 10 - 4t$
    $t \leq 8$

13–15. Express *y* as a function of *x* using the equations in Exercises 10–12. State any restriction on *x*.

· · · · · · · · · · · · · · · · · · · ·
Spiral Review

16. **Open-ended** Give an example of two variables whose relationship shows correlation but not causation. *(Section 2.5)*

The population of Harbin, China has been growing at the rate of 1.5% per year. In 1995, its population was about 2.7 million. *(Section 1.2)*

17. Write an equation to model the population of Harbin, China.

18. What will be the population of Harbin in the year 2000?

Write a point-slope equation of the line through each pair of points. *(Section 2.3)*

19. (2, 5) and (−7, 7)

20. (3, 4) and (5, 18)

21. $\left(-4, -\frac{1}{2}\right)$ and $\left(-6, \frac{5}{2}\right)$

Chapter 2 Review

Complete these exercises for a review of Chapter 2. If you have difficulty with a particular problem, review the indicated section.

For each graph, tell whether *y* varies directly with *x*. If so, give an equation for the graph. *(Section 2.1)*

1.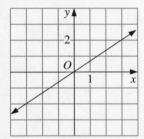

2.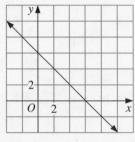

3.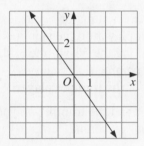

For each equation in Exercises 4–6, tell whether *y* varies directly with *x*. If so, graph the equation. *(Section 2.1)*

4. $y = 2.1x$

5. $x = 3y$

6. $y + 2 = 0.5x$

7. State the slope and *y*-intercept of the line with the equation $y = 2 - \frac{4}{5}x$. *(Section 2.2)*

8. Find the slope-intercept equation of the line through the points $(5, 0)$ and $(-2, 8.4)$. *(Section 2.2)*

9. Graph the equation found in Exercise 8. *(Section 2.2)*

For Exercises 10 and 11, write a point-slope equation of the line passing through the given points. *(Section 2.3)*

10. $(-1, 5)$ and $(3, 1)$

11. $(2, -3)$ and $(5, 5)$

12. Find the domain and range of the function $f(x) = -\frac{2}{5}x + 2$ for $x \le 5$. *(Section 2.3)*

The table shows the number of stories in several tall buildings in Chicago and the height of each building. Use this table for Exercises 13–15.

Building	Number of stories	Height (ft)
Sears Tower	110	1454
Amoco	82	1136
900 N. Michigan	66	871
Three First National Plaza	57	775
IBM Plaza	52	695
CNA Plaza	44	600

13. **Writing** Is the relationship between the building height and the number of stories a direct variation? Explain your reasoning. *(Section 2.1)*

14. Make a scatter plot of the data and draw a line of fit. Plot the number of stories along the horizontal axis and building height along the vertical. Give an equation for the line of fit. *(Section 2.4)*

15. **Technology** Use a calculator or statistical software to find an equation of the least-squares line and the correlation coefficient for the data in the table. *(Sections 2.4 and 2.5)*

For Exercises 16–18, match each scatter plot to one of these following values of *r*: –0.9, 0.9, –0.5, 0.3. *(Section 2.6)*

16.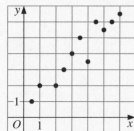

17.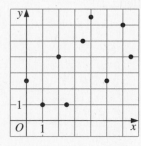

18.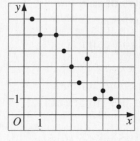

19. Graph the pair of parametric equations at the right for the given restriction on *t*. *(Section 2.6)*

$$x = 3 - 0.5t$$
$$y = 4t$$
$$0 \le t \le 8$$

20. For the equations in Exercise 19, express *y* as a function of *x*. State any restriction on *x*. *(Section 2.6)*

21. **Open-ended** Write a pair of parametric equations, with restrictions on *t*, that will describe a path from the point (0, 1) to the point (5, 6). *(Section 2.6)*

SPIRAL REVIEW   **Chapters 1–2**

For Exercises 1–3, simplify each expression.

1. $4 + 2 \times 6 - 1$ 2. $4(3 + 2)^2$ 3. $200 \div 100 \div 4$

4. **Writing** Matrices *A* and *B* are two 2×3 matrices such that their sum is the

matrix $\begin{bmatrix} 0 & 0 & 0 \\ 0 & 0 & 0 \end{bmatrix}$. What can you say about matrices *A* and *B*?

The pupils of the human eye dilate and constrict in response to the amount of light available. Biologists have found that the ability of an eye to dilate and constrict its pupil size varies with age as indicated in the table at the right.

Age (years)	Range of pupil diameter (mm)
20	3.3
30	2.7
40	2.1
50	1.5

5. Find the difference in the range of pupil diameter in each 10-year period. What is the average change in pupil range per 10-year period? per year?

6. Without using technology, give the correlation coefficient of the pupil range data.

Exponential Growth and Decay

Learn how to . . .

- describe growth and decay using tables of data and equations

• **So you can . . .**

use exponential equations to model real-life situations

Application

A large community chorus establishes a "telephone tree" to quickly inform its members of changes in the rehearsal schedule. First, the president calls three members. Next, these three members each call three more members. Then, these nine members each call three more members. This process continues until all the members are called.

Terms to Know

Exponential growth (p. 96)
    growth that can be modeled by an equation of the form
    $y = ab^x$, where $a > 0$ and $b > 1$

Example / Illustration

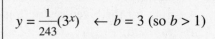

$$y = \frac{1}{243}(3^x) \quad \leftarrow b = 3 \text{ (so } b > 1)$$

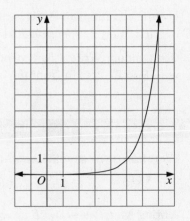

Exponential decay (p. 96)
  decay that can be modeled by an equation of the form
  $y = ab^x$, where $a > 0$ and $0 < b < 1$

$$y = 81\left(\frac{1}{3}\right)^x \quad \leftarrow \quad b = \left(\frac{1}{3}\right)$$

(so $0 < b < 1$)

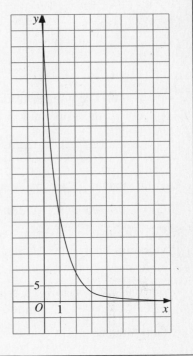

UNDERSTANDING THE MAIN IDEAS

The following properties of exponents will be used throughout this chapter.

For any $b > 0$ and positive integers m and n:

Product Property: $b^m \cdot b^n = b^{m+n}$

Quotient Property: $\dfrac{b^m}{b^n} = b^{m-n}$

Power Property: $(b^m)^n = b^{mn}$

Example 1

Refer to the situation described in the Application.

a. Make a table showing the number of phone calls made at each of the first four levels of the "telephone tree."

b. Write an equation that models the number of phone calls made at each level of the "telephone tree."

■ Solution ■

a.

Level	Number of calls	
1	3	← The president calls three members.
2	$3 \times 3 = 9$	← Each of the 3 called in level 1 makes 3 calls.
3	$9 \times 3 = 27$	← Each of the 9 called in level 2 makes 3 calls.
4	$27 \times 3 = 81$	← Each of the 27 called in level 3 makes 3 calls.

b. The number of calls, C, made at the nth level of the "telephone tree" is given by the equation $C = 3^n$.

1. How many calls are made in level 5 of the telephone tree?

2. Graph the function found in part (b) of Example 1 for $n = 0, 1, 2, ..., 5$.

3. Suppose that the telephone tree is changed so that both the president and the vice-president of the chorus make 3 calls each in the first round. Make a table showing the number of calls in each of the first four levels.

4. For the situation in Exercise 3, write an equation that models the number of calls, C, made in the nth level of the telephone tree.

5. Graph the equation you wrote for Exercise 4.

Equations for exponential growth or decay

The telephone tree discussed in Example 1 describes a situation involving exponential growth. Some situations involve exponential decay; that is, they involve a quantity that is decreasing at an exponential rate. The equations that model exponential growth and exponential decay have the same general form,

$$y = ab^x$$

where y = the amount after x events (seconds, minutes, years, and so on), a = the original amount, b = the growth or decay factor, and x = the number of events. (*Note*: In this section, the value of a is always positive.) If $0 < b < 1$, then the equation is a model of exponential decay—the value of y is decreasing exponentially. If $b > 1$, then the equation is a model of exponential growth—the value of y is increasing exponentially.

Example 2

A land developer has an 800–acre parcel of land to sell. In the first week, the developer sells 200 acres. During the second week, the developer sells one–fourth as many acres as the previous week, 50 acres. Suppose this pattern continues and that each week the developer sells one–fourth as much land as the previous week. Write an equation modeling the amount of land sold each week.

6. Make a table showing the amount of land sold in each of the first 5 weeks.

7. How does the equation change if instead of starting with 800 acres of land to sell, the developer has 400 acres?

8. A 5-gram colony of bacteria grows at a pace such that it doubles in mass every day. Write an exponential growth equation modeling the mass B of the colony after n days.

Evaluate each expression when $x = 3$.

9. $0.25(4^x)$

10. $243\left(\frac{2}{3}\right)^{x+1}$

11. $1000(1.01)^x$

Tell whether each equation models *exponential growth*, *exponential decay*, or *neither*.

12. $y = 7(0.8^x)$

13. $y = 0.8x + 7$

14. $y = 0.8(7^x)$

· · · · · · · · · · · · · · · · · · ·
Spiral Review

For each pair of parametric equations, express y as a function of x.
(Section 2.6)

15. $x = 4t$
    $y = 0.25(t + 8)$

16. $x = t - 5$
    $y = t + 4$

17. $x = 3t + 8$
    $y = 2t$

For Exercises 18–20, tell whether each equation shows direct variation.
(Section 2.1)

18. $y = 0.8x$

19. $3y + 6 = x + 6$

20. $y = \frac{1}{3}x - 5$

21. Write a linear equation that gives the number of ounces, o, that are in p pounds. Is this an example of a direct variation? *(Section 2.1)*

3.2 Negative and Rational Exponents

Learn how to . . .

- evaluate expressions that use negative and rational exponents

So you can . . .

- describe situations involving continuous exponential growth or decay

Application

Negative and rational exponents are used to describe the frequencies of musical notes. For example, the frequency of the note middle C is $220(2)^{3/12}$ Hz. (The abbreviation Hz stands for Hertz, a unit used when measuring the frequency of a sound.) The C note just below middle C has a frequency of $220(2)^{-9/12}$ Hz.

UNDERSTANDING THE MAIN IDEAS

Negative exponents and rational exponents

For any base $b > 0$ and any positive integer n:

$$b^{-n} = \frac{1}{b^n}$$

For any base $b > 0$ and any positive integer n:

$$b^{1/n} = c \text{ if and only if } b = c^n$$

Example 1

Use the properties of exponents to simplify each expression.

  a. 6^{-2} **b.** $25(0.4)^{-3}$ **c.** $125^{2/3}$ **d.** $36^{1/8} \cdot 36^{3/8}$ **e.** $\dfrac{8^{1/3}}{8^{2/3}}$

■ Solution ■

  a. $6^{-2} = \dfrac{1}{6^2} = \dfrac{1}{36}$ $\leftarrow b^{-n} = \dfrac{1}{b^n}$

(Solution continues on next page.)

■ Solution ■ *(continued)*

b. $25(0.4)^{-3} = 25\left(\dfrac{1}{0.4}\right)^3$ ← The exponent applies only to the factor $\dfrac{1}{0.4}$.
It is not applied to the factor 25.

$\qquad\qquad = 25\left(\dfrac{1}{0.4^3}\right)$

$\qquad\qquad = \dfrac{25}{0.4^3}$

$\qquad\qquad = \dfrac{25}{0.064}$

$\qquad\qquad = 390.625$

c. $125^{2/3} = (5^3)^{2/3}$ ← $125 = 5^3$

$\qquad\qquad = 5^{3 \,\cdot\, 2/3}$ ← $(b^m)^n = b^{mn}$

$\qquad\qquad = 5^2$

$\qquad\qquad = 25$

d. $36^{1/8} \cdot 36^{3/8} = 36^{1/8 \,+\, 3/8}$ ← $b^m b^n = b^{m+n}$

$\qquad\qquad\qquad = 36^{1/2}$

$\qquad\qquad\qquad = 6$ ← $36^{1/2} = 6$ because $6^2 = 36$

e. $\dfrac{8^{1/3}}{8^{2/3}} = 8^{1/3 \,-\, 2/3}$ ← $\dfrac{b^m}{b^n} = b^{m-n}$

$\qquad\quad = 8^{-1/3}$

$\qquad\quad = \left(\dfrac{1}{8}\right)^{1/3}$ ← $b^{-n} = \dfrac{1}{b^n}$

$\qquad\quad = \dfrac{1}{8^{1/3}}$

$\qquad\quad = \dfrac{1}{2}$ ← $8^{1/3} = 2$ because $2^3 = 8$

Simplify using the properties of exponents.

1. 4^{-2}

2. $500(2.5)^{-4}$

3. $3^3 \cdot 3^{-4}$

4. $3x^{20}x^{-16}$

5. $9^{1/2}$

6. $\dfrac{4^2}{4^{2.5}}$

7. $17^{-0.6} \cdot 17^{0.6}$

8. $27^{2/9} \cdot 27^{1/9}$

9. $4^{1/4} \cdot 2^{1/2}$

Example 2

An auction house estimates that the value of an oil painting has been increasing at the rate of 12% per year. They set the current value of the painting at $300,000.

a. Write an exponential function relating the value of the painting to the number of years from now.

b. How much was the painting worth 20 years ago?

c. How much will the painting be worth in 18 months?

■ Solution ■

a. Let x = the number of years from now and let y = the value of the painting. Since the value of the painting is *increasing* by 12% each year, the *growth factor* is 100% + 12% = 112%, or 1.12. The current value of the painting is $300,000.

$y = ab^x$ ← Substitute 300,000 for a and 1.12 for b.

$y = 300,000(1.12)^x$

b. Use the equation found in part (a), with $x = -20$.

$y = 300,000(1.12)^{-20}$

$\approx 31,100$

The value of the painting 20 years ago was about $31,100.

c. Since 18 mo = 1.5 yr, use the equation found in part (a) with $x = 1.5$.

$y = 300,000(1.12)^{1.5}$

$\approx 356,000$

The value of the painting 18 months from now will be about $356,000.

The value of a truck decreases 18% each year. The current value of the truck is $52,000. Use this information for Exercises 10–12.

10. Write an exponential function that models the value, y, of the truck t years from now. (*Hint*: The decay factor is 100% – 18% = 82%, or 0.82.)

11. Use the function you wrote for Exercise 10 to find the value of the truck 3 years from now.

12. Use the function you wrote for Exercise 10 to find the value of the truck 30 months ago.

13. The function $y = 10.105(1.03)^t$, where t represents the number of years before or after 1995, models the estimated population (in millions) of the city of Delhi, India. What was the population of Delhi in 1992? If population growth continues in this pattern, what will be the population of Delhi in the year 2005?

14. Mathematics Journal For the exponential function $y = ab^x$, explain how the value of b relates to the rate of growth or decay.

Study Guide, ALGEBRA 2: EXPLORATIONS AND APPLICATIONS

Evaluate each expression. *(Section 3.1)*

15. $6(3^3)$

16. $1.6(0.2)^3$

17. $3.6(2^{12} \cdot 2^{-12})$

Compare each graph to the graph of $y = 5 - \frac{1}{2}x$. *(Section 2.2)*

18. $y = -2 - \frac{1}{2}x$

19. $y = 5 + \frac{1}{2}x$

20. $y = 5 - \frac{1}{2}(x + 3)$

21. Design and carry out a simulation to estimate how many people are needed on average before finding one who was born on a Sunday. *(Section 1.5)*

63

Graphs of Exponential Functions

GOAL

Learn how to . . .

- draw graphs of exponential functions

- interpret how different values of *a* and *b* affect the graph of $y = ab^x$

So you can . . .

- determine the doubling time or the half-life of a quantity

Application

The element carbon-14 is radioactive and decays exponentially over time. Its half-life is 5730 years. Archaeologists use this information to determine the age of ancient objects.

Terms to Know

Example / Illustration

Doubling time (p. 108)
   when something is growing exponentially, the time it takes for an initial amount to double

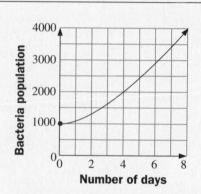

The doubling time for this population of bacteria is 4 days, because in 4 days the population doubles from 1000 to 2000.

Half-life (p. 109)
   when something is decaying exponentially, the time it takes
   for an initial amount to be halved

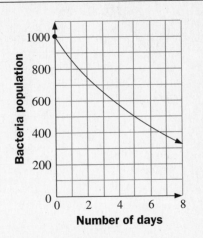

The half-life of this bacteria
population is 5 days, because in
5 days the population decreases
from 1000 to 500.

UNDERSTANDING THE MAIN IDEAS

The values of a and b in the exponential function $y = ab^x$ affect the graph of the
function. When $a > 0$, the entire graph is above the x-axis. When $a < 0$, the
entire graph is below the x-axis. For the graphs of all exponential functions of
the form $y = ab^x$, the y-intercept is a.

Example 1

Match the graphs with their equations.

I. 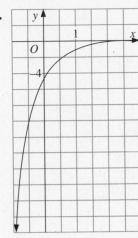   II.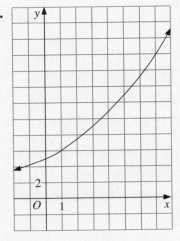

III.

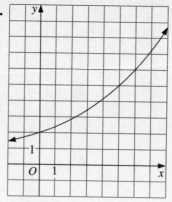

IV.

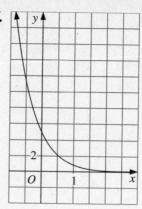

a. $y = 5(1.2)^x$

c. $y = 5(0.2)^x$

b. $y = 2(1.2)^x$

d. $y = -5(0.2)^x$

▪ Solution ▪

a. Since $a = 5$ in the equation, the y-intercept of the graph is 5. There are two possible choices, graphs II and IV. Also, since $a > 0$ and $b > 1$, the function represents exponential growth. Therefore, the correct choice is graph II.

b. Since $a = 2$ in the equation, the y-intercept of the graph is 2. Also, since $a > 0$ and $b > 1$, the function represents exponential growth. Therefore, the correct choice is graph III.

c. Since $a = 5$ in the equation, the y-intercept of the graph is 5. Also, since $a > 0$ and $0 < b < 1$, the function represents exponential decay. Therefore, the correct choice is graph IV.

d. Since $a = -5$ in the equation, the y-intercept of the graph is -5. So the correct choice is graph I. (*Note:* The graph of this function is the graph of $y = 5(0.2)^x$ reflected over the x-axis.)

1. **Technology** Use a graphing calculator or graphing software to compare the graph of $y = 5(1.2)^x$ to the graphs of $y = 5(1.4)^x$ and $y = 5(1.1)^x$.

2. How do the graphs of $y = 5(1.2)^x$ and $y = -5(1.2)^x$ compare?

3. How do the graphs of $y = 5(0.2)^x$ and $y = 3(0.2)^x$ compare?

For each function in Exercises 4–6, do the following:

a. Find the y-intercept of the graph.

b. Tell whether the graph represents *exponential growth* or *exponential decay*.

c. Sketch the graph.

4. $y = 10(0.8)^x$ 5. $y = 5^x$ 6. $y = 20(1.1)^x$

7. **Writing** Explain why a is the y-intercept of the graph of any function of the form $y = ab^x$.

Doubling time and half-life

When a quantity is growing exponentially, the time it takes for an initial amount to double is called the *doubling time*. Recall that an exponential function represents *growth* when a is positive and b is greater than 1. Such a growth function can always be rewritten in this form:

$$y = a(2)^{x/d}, \text{ where } a = \text{the initial amount and } d = \text{the doubling time.}$$

When a quantity is decaying exponentially, the time it takes for there to be just half of the initial amount is called the *half-life*. Recall that an exponential function represents *decay* when a is positive and b is greater than 0 and less than 1. Such a decay function can be rewritten in this form:

$$y = a\left(\frac{1}{2}\right)^{x/h}, \text{ where } a = \text{the initial amount and } h = \text{the half-life.}$$

Example 2

a. Radioactive iodine is used in a medical test to determine the health of a patient's thyroid gland. Radioactive iodine has a half-life of 8.1 days. Write an equation that gives the amount, y, of radioactive iodine that is present x days after an initial dose, A_0, was administered to a patient.

b. Write the equation found in part (a) in the form $y = ab^x$. Use this equation to find the rate of decay per day for radioactive iodine.

■ Solution ■

a. Since the situation involves finding the half-life, the equation has the form

$$y = a\left(\frac{1}{2}\right)^{x/h}, \text{ with } a = A_0 \text{ and } h = 8.1.$$

$$y = A_0\left(\frac{1}{2}\right)^{x/8.1}$$

b. $y = A_0\left(\frac{1}{2}\right)^{x/8.1}$

$$= A_0\left(\left(\frac{1}{2}\right)^{1/8.1}\right)^x \qquad \leftarrow b^{mn} = (b^m)^n$$

$$\approx A_0(0.918)^x \qquad \leftarrow \text{Use a calculator: } \left(\frac{1}{2}\right)^{1/8.1} \approx 0.918.$$

So the decay *factor* is about 0.918. This means that every day, the amount of radioactive iodine is about 91.8% of the amount the previous day. The decay *rate* is $1 - 0.918 \approx 0.082$, or about 8.2%.

8. Substitute 8.1 for x in the equations found in parts (a) and (b) of Example 2 to find the fraction of A_0 left after 8.1 days.

9. How many days will it take for the amount of radioactive iodine remaining to be $\frac{1}{4}A_0$? $\frac{1}{8}A_0$?

10. Write an equation for the amount y, of carbon-14 left in a sample after x years. Use the information in the Application at the beginning of this section. Use A_0 for the initial amount of carbon-14 present.

11. Find the decay factor and the decay rate per year for carbon-14. Round your answers to five decimal places.

A school principal in a growing community expects the school population to grow exponentially and to double in 20 years. There are 355 students in the school this year.

12. Write an equation in the form $y = a(2)^{x/d}$ relating the number of years, x, to the number of students, y.

13. By what percent is the school population growing each year?

·····················
Spiral Review

Simplify. *(Section 3.2)*

14. $16^{1/4}$ 15. 2^{-4} 16. $25^{3/2}$

The equation $P = 27.54(1.011)^x$ models the population P of Tokyo-Yokohama, Japan in millions, with x = years since 1992. Find the value of P for each year. *(Section 3.2)*

17. 1995 18. 1990 19. 1982

20. What is the doubling time for the population of Tokyo-Yokohama? *(Section 3.3)*

The table shows the maximum suggested weight of adults, ages 19 to 34. *(Section 2.4)*

21. Graph the data and draw a line of fit.

22. Find an equation for the line of fit.

23. Make a prediction about the maximum suggested weight of a 22 year old man who is 68 in. tall.

Height (in.)	Maximum suggested weight (lb)
60	128
65	150
70	174
75	200

The Number *e*

Application

The number *e* can be used as the base of any exponential function that is a model for a quantity that grows or decays *continuously*. Suppose a chemical reaction consumes water continuously at the rate of 1.5% of the available water per hour. The amount of water, *y*, after *t* hours is modeled by the equation $y = A_0 e^{-0.015t}$, where A_0 is the initial amount of water present.

Terms to Know	**Example / Illustration**
e (p. 116) the value approached by the expression $\left(1 + \dfrac{1}{n}\right)^n$ as *n* increases ($e = 2.718281828\ldots$)	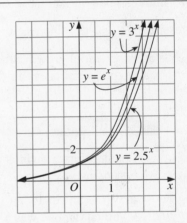 The graph of $y = e^x$ lies between the graphs of $y = 2.5^x$ and $y = 3^x$ because *e* is between 2.5 and 3 ($2.5 < e < 3$).

Terms to Know	Example / Illustration

Logistic growth (p. 118)
a kind of limited growth that begins with rapid growth but eventually levels off due to some limiting factor(s)
(Equations for logistic growth are often written using e.)

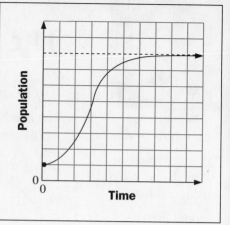

UNDERSTANDING THE MAIN IDEAS

Exponential functions involving e

When interest on a bank account is compounded continuously, the formula $A = Pe^{rt}$ is used to compute the amount of money in the account. The letter e in the formula is not a variable, but rather the number approximated by the expression $\left(1 + \dfrac{1}{n}\right)^n$ as n increases ($e \approx 2.718$). Because $e > 1$, the function $y = e^x$ is an exponential growth function. Also, because $e^{-1} < 1$, the function $y = e^{-x}$ is an exponential decay function.

Example

A bank offers a savings account that pays 4% annual interest, compounded continuously.

a. Write an equation that models the amount, A, in the account after t years if the initial deposit was $5000 and no additional money was ever deposited.

b. Use the equation in part (a) to find the amount of money in the account after 2.5 years.

c. How long will it take for the amount of money in the account to double?

d. Write an equation using base 2 rather than base e for the amount of money in the account after t years.

Study Guide, ALGEBRA 2: EXPLORATIONS AND APPLICATIONS

▪ Solution ▪

a. Use the continuous compound interest formula $A = Pe^{rt}$.

$A = 5000e^{0.04t}$ ← Substitute 5000 for P and $4\% = 0.04$ for r.

(*Note*: In the continuous compound interest formula, r is the interest rate (growth rate) expressed as a decimal. Do not use the growth factor.)

b. Substitute 2.5 for t in the equation from part (a).

$A = 5000e^{0.04(2.5)}$

$= 5000e^{0.1}$

≈ 5525.85459

The amount of money in the account after 2.5 years is $5525.85.

c. When the amount of money in the account has doubled, the value of A in the equation from part (a) is 10,000.

On a graphing calculator, graph these two equations on the same set of axes:

$y_1 = 5000e^{0.04x}$ and $y_2 = 10,000$

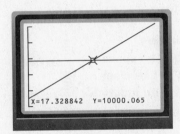

The value of x at the point where the two graphs intersect is the number of years it takes for the amount of money in the account to double. The graphs intersect at $x \approx 17.33$ and $y = 10,000$.

The doubling time is approximately $17\frac{1}{3}$ years (or 17 years 4 months).

d. Recall from Section 3.3 that a growth function can always be rewritten in the form $y = a(2)^{x/d}$, where a = the initial amount and d = the doubling time. (Notice that the base in this function is 2.) In this case, 5000 is the initial amount and 17.33 is the doubling time.

$A = 5000(2)^{t/17.33}$

---

1. Technology Use a graphing calculator to graph the equations found in parts (a) and (d) of Example 1 on the same set of axes. How do they compare? Explain.

2. By what percent does the account in Example 1 grow in a year? (This percent is called the *effective annual yield*.)

Find the value of an investment of $1000 after 5 years in each of these situations.

3. an account paying 8.7% interest, compounded monthly

4. an account paying 6.1% interest, compounded daily

5. an account paying 4.2% interest, compounded continuously

Find the value of $\left(1 + \dfrac{1}{n}\right)^n$ **for each value of _n_. Round each answer to six decimal places.**

6. 999   7. 9999   8. 999,999

Radioactive phosphorus decays continuously at the rate of 4.95% per day. This means that an initial amount, A_0, of radioactive phosphorus decays after _t_ days according to the formula $A = A_0 e^{-0.0495t}$.

9. The mass of an initial sample of radioactive phosphorus is 5 g. How much of the sample is left after 20 days?

10. **Writing** Explain why the exponent in the formula is negative.

11. Find the half-life of radioactive phosphorus.

12. Rewrite the formula for _A_ using $\dfrac{1}{2}$ as the base instead of _e_.

A population of bacteria was being studied in a laboratory. The population grew rapidly at first, but because food and space were limited, the population eventually leveled off. For Exercises 13–15, use the graph at the right and the logistic growth function
$$y = \dfrac{6000}{1 + 11e^{-0.05x}}.$$

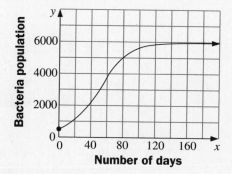

13. What was the initial population of the bacteria?

14. How long did it take for the population to double? to quadruple? to be 8 times the initial population?

15. At what population did the bacteria eventually stabilize?

16. **Technology** Use a graphing calculator to graph the logistic growth function
$y = \dfrac{50}{1 + 9e^{-0.2x}}$. What is the initial value of _y_? What happens to the value of _y_ as the value of _x_ increases?

· · · · · · · · · · · · · · · · · · · ·
Spiral Review

For Exercises 17 and 18 refer to the graph of $y = 100b^x$. *(Section 3.3)*

17. **Open-ended** Give three possible values of _b_ for which the graph would show exponential decay. Which of the three values yields the graph with the slowest decay?

18. Let _x_ = the number of hours. What value of _b_ would show a growth rate of 13% per hour?

Exercises 19–21 refer to the table which shows the affect of advertising on the sales of jeans. *(Section 2.4)*

Amount spent on advertising (thousands of dollars)	Pairs of jeans sold (thousands)
40	28
100	65
180	120
220	138

19. Show that the data have a linear relationship by making a scatter plot and drawing a line of fit.

20. Technology Use a graphing calculator or statistical software to find an equation of the least-squares line that fits the data.

21. Predict the number of pairs of jeans sold if $150,000 is spent on advertising.

Multiply. *(Section 1.4)*

22. $\begin{bmatrix} 2 & 0 \\ -1 & 6 \end{bmatrix}\begin{bmatrix} 0 & 1 & 4 \\ 3 & 6 & 5 \end{bmatrix}$

23. $\begin{bmatrix} 5 & 0 & 1 \\ 1 & 0 & 2 \\ 1 & 0 & 0 \end{bmatrix}\begin{bmatrix} 3 \\ 5 \\ 9 \end{bmatrix}$

Fitting Exponential Functions to Data

Learn how to . . .

- write exponential functions that fit sets of data

So you can . . .

- make predictions about exponential growth and decay situations

Application

Economists use data about the national debt to write an exponential model of its growth and then use the model to predict the national debt in future years.

UNDERSTANDING THE MAIN IDEAS

Writing exponential functions to fit data

When a set of data appears to grow or decay exponentially, you can try to find an exponential equation that is a "good fit." To do so, you need to find appropriate values of a and b for an equation of the form $y = ab^x$. You can use a graphing calculator or statistical software to find an exponential equation that is a good fit for the data.

Example

The table at the right shows the average gross weekly earnings of a worker in the manufacturing industry in the United States for the years 1975 and 1990. Assume that manufacturing wages have grown exponentially since 1900. Write an equation that estimates the average gross weekly earnings of a manufacturing worker x years after 1900.

Manufacturing Wages	
Year	Average gross weekly earnings
1975	$190
1990	$442

Study Guide, ALGEBRA 2: EXPLORATIONS AND APPLICATIONS

▪ Solution ▪

The exponential equation will be in the form $y = ab^x$. The graph of the equation must pass through the two points $(x, y) = (75, 190)$ and $(x, y) = (90, 442)$.

Step 1: Substitute each pair of values into the equation $y = ab^x$.

$$190 = ab^{75} \qquad \text{and} \qquad 442 = ab^{90}$$

Step 2: Since the variable a has the same exponent in both equations while the exponents of b are different, divide the two equations and solve the resulting equation for b. (*Note*: We know $a \neq 0$ and $b \neq 0$.)

$$\frac{442}{190} = \frac{ab^{90}}{ab^{75}}$$

$$2.33 \approx \frac{b^{90}}{b^{75}} \qquad \leftarrow \text{Cancel } a \text{ in the numerator and denominator.}$$

$$2.33 \approx b^{15} \qquad \leftarrow \frac{b^{90}}{b^{75}} = b^{90-75}$$

$$(2.33)^{1/15} \approx b^{(15 \, \cdot \, 1/15)} \qquad \leftarrow \text{Use a calculator.}$$

$$1.06 \approx b$$

Step 3: Now solve for a by substituting the value of b into either equation from Step 1.

$$190 = a(1.06)^{75} \qquad \leftarrow \text{Use a calculator.}$$

$$190 \approx a(79.1)$$

$$\frac{190}{79.1} \approx a \qquad \leftarrow \text{Use a calculator.}$$

$$2.40 \approx a$$

An equation that describes the average gross weekly earnings of a manufacturing worker in the United States is $y = 2.40(1.06)^x$.

---

1. Use the equation found in the Example to estimate the average gross weekly earnings for a worker in 1970, 1985, and 1995.

2. How fast did manufacturing earnings grow from 1975 to 1990?

3. Write an equation modeling manufacturing earnings in the form $y = ab^x$, where $x =$ the number of years since 1975.

Write an exponential function whose graph passes through each pair of points.

4. $(4, 18)$ and $(6, 3)$ 5. $(3, -4)$ and $(9, -256)$ 6. $(0, 50)$ and $(12, 300)$

From the time a patient takes a medication for high blood pressure, the amount of the medicine in the bloodstream decays exponentially. To determine the rate of decay for a particular patient, a doctor gives the patient a 75 mg dosage. A sample of the patient's blood is taken 5 h later and the doctor determines that 13.5 mg of the medicine remain in the bloodstream.

7. Write an exponential function modeling the amount of medicine in the bloodstream x hours after it is taken.

8. What is the half-life of this medicine?

9. What is the rate of decay of this medicine in the bloodstream? Give your answer in percent per hour.

10. Doctors consider this medicine effective if there is more than 1.2 mg in the bloodstream. Is the medicine still effective after 12 h?

The information in the table at the right is more complete than the information presented in the Example.

11. How do the values given by the function found in the Example compare to the actual data in the table?

12. **Technology** Use a graphing calculator or statistical software to find an exponential equation that gives a good fit. Compare your result with the equation found in the Example.

Manufacturing Wages	
Year	Average gross weekly earnings
1970	$134
1975	$190
1980	$289
1985	$386
1990	$442

The table at the right shows the amount of money spent by the United States government for interest on the public debt. Use this information for Exercises 13–16.

13. Make a scatter plot of the data in the table.

14. **Open-ended** Pick any two years and write an exponential equation that models the growth of interest payments on the public debt of the United States. Let x = the number of years since 1960.

15. Check the equation you found in Exercise 14 by comparing its predictions to the data for the other years in the table.

16. **Technology** Enter the data into a graphing calculator or statistical software to find an exponential equation that gives a good fit. Let x = the number of years since 1960. Compare the result to the equation you found in Exercise 14. What is the annual rate of growth of the interest payments suggested by each equation?

Year	Interest (millions of dollars)
1960	9,180
1970	19,304
1975	32,665
1980	74,860
1985	178,945
1990	768,725
1993	918,663

17. **Mathematics Journal** Describe some situations where it is useful to have an exponential function modeling data showing growth or decay.

The growth of the population of a city is modeled by the equation
$P(t) = P_0 e^{0.021t}$, **where** $P(t)$ **is the population in thousands after** t **years and**
P_0 **is the population when** $t = 0$. *(Section 3.4)*

18. In January of 1930, the population of the city was 2.5 thousand. Write a function for the population after 1930.

19. What was the population of the city in January of 1942?

20. What was the population of the city in January of 1920?

21. How would you find the population of the city in May of 1954?

For Exercises 22 and 23, use the equation $y = -6 + 4x$. *(Section 2.2)*

22. Find y when $x = -2$.   **23.** Find x when $y = 12$.

24. Graph the parametric equations $x = 25 + 0.3t$ and $y = 1.2t - 10$ for $t \geq 0$. *(Section 2.6)*

25. For the parametric equations in Exercise 24, express y as a function of x. State any restriction on x. *(Section 2.6)*

Chapter 3 Review ···············

Complete these exercises for a review of Chapter 3. If you have difficulty with a particular problem, review the indicated section.

For Exercises 1–3, evaluate each expression when $x = 4$. *(Section 3.1)*

1. $3(4)^x$

2. $20\left(\dfrac{1}{2}\right)^{x+1}$

3. $500(1.2)^x$

4. Open-ended Give an example of an equation representing each of the following: exponential growth, linear growth, exponential decay, linear decay. *(Section 3.1)*

Simplify using the properties of exponents. *(Section 3.2)*

5. $9^{3/2} \cdot 9^{-4}$

6. $\dfrac{p^{12}}{p^5}$

7. $(15^{-1/3})^6$

The equation $P = 3000(1.019)^t$ models the size of a population of bacteria, where t represents the number of hours past 12 noon.

8. Estimate the size of the bacteria population at 10 A.M., 3 P.M., and 9:15 P.M. *(Section 3.2)*

9. What is the doubling time of the population of bacteria? *(Section 3.3)*

10. By what percent is the bacteria population growing each hour? each day? *(Section 3.3)*

The half-life of the isotope tritium is 12.5 years. Suppose you start with a 10-gram sample. Use this information for Exercises 11 and 12. *(Section 3.3)*

11. Write an equation modeling the amount of tritium that will be left in t years.

12. For how long will there be more than 7 g of tritium?

13. Writing Describe the shape of graphs representing exponential growth and exponential decay. *(Section 3.3)*

The logistic growth function $y = \dfrac{67}{1 + 1.484e^{-0.239t}}$, where t represents the number of years since 1980, models the percent of United States households with cable television. *(Section 3.4)*

14. What percent of U.S. households had cable television in 1980? in 1984? in 1977?

15. Technology Graph the function using a graphing calculator or graphing software. Estimate the year in which the percent of households with cable television reached 50%.

16. Using this model, at what value will the percent of U.S. households with cable television eventually stabilize?

Find the value of an investment of $15,000 after 8 years if the interest rate is 6.8% and the interest is compounded as specified. *(Section 3.4)*

17. monthly   **18.** annually   **19.** continuously

Write an exponential function whose graph passes through each pair of points. *(Section 3.5)*

20. (3, 90) and (5, 100)   **21.** (0, 18) and (15, 3)

The value of a piece of manufacturing equipment decreases exponentially from the time of purchase. This decrease is called *depreciation.* **The equipment was purchased for $25,000 and 3 years later it is valued at $22,000.** *(Section 3.5)*

22. Write an exponential function modeling the value of the equipment x years after it was purchased.

23. Use the model to estimate the value of the equipment 10 years after it was purchased.

SPIRAL REVIEW   Chapters 1–3

Greg Power can bowl a "strike" about 2 out of every 3 attempts. He would like to break his team record by bowling 4 strikes in a row. He wonders how long he will have to try.

1. Describe how you could use a die to simulate this situation.

2. The data from six trials of a simulation are shown in the table below. In the table, "S" represents "strike" and "N" represents "not a strike." Use the data to estimate the number of times Greg will have to try before bowling four strikes in a row.

Trial	Result
1	S, S, N, S, S, S, S
2	N, S, N, S, S, S, S
3	S, S, N, N, N, S, S, S, N, N, S, S, N, N, N, S, S, S, S
4	N, S, S, S, S
5	S, N, S, S, S, S
6	N, N, S, N, S, S, S, N, N, S, N, S, S, S, S

Tell whether each of these equations represents *linear growth, linear decay, exponential growth,* **or** *exponential decay.*

3. $y = 5\left(\frac{1}{3}\right)^x$ **4.** $y = 5 + \frac{1}{3}x$ **5.** $y = \left(\frac{1}{3}\right)5^x$ **6.** $y = \frac{1}{3} - 5x$

Tiffany and Bryanna earn money baby-sitting. Tiffany charges $5 for every job, plus $1.50 per hour. Bryanna charges $2.50 per hour.

7. Write linear equations modeling the amount each baby sitter makes on a job. Let x represent the number of hours worked.

8. Which baby sitter's earnings vary directly with her hours?

Using Inverses of Linear Functions

Learn how to . . .

- graph and find equations for inverses of linear functions

So you can . . .

- solve problems

Application

The formula that converts Celsius to Fahrenheit is the inverse of the formula that converts Fahrenheit to Celsius.

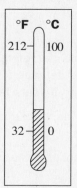

Terms to Know

Example / Illustration

Inverse of a function (p. 142) a function that satisfies the property $f^{-1}(b) = a$ if and only if $f(a) = b$ (The inverse of a function f is denoted f^{-1}, which is read as "f inverse.")	The functions $F(x) = \dfrac{9}{5}x + 32$ and $C(x) = \dfrac{5}{9}(x - 32)$ are inverses. For example, $F(100) = 212$ and $C(212) = 100$, and $F(0) = 32$ and $C(32) = 0$.

UNDERSTANDING THE MAIN IDEAS

Graphing functions and their inverses

If a function has an inverse, the graph of the inverse is the reflection of the graph of the function across the line $y = x$.

Example 1

Let $g(x) = 2x - 4$. Graph $y = g(x)$ and $y = g^{-1}(x)$ in the same coordinate plane.

■ Solution ■

Note that the function $g(x)$ is a linear function given in slope-intercept form.

Step 1: Graph $y = g(x)$. The graph is a line with slope 2 and y-intercept -4, so it passes through the points $(0, -4)$ and $(2, 0)$.

Step 2: Draw the line $y = x$.

Step 3: Graph $y = g^{-1}(x)$ by reflecting the graph of $y = g(x)$ over the line $y = x$. To do this, reverse the coordinates of any points on the graph of $y = g(x)$ to obtain the coordinates of points on the graph of $y = g^{-1}(x)$. Since the points $(0, -4)$ and $(2, 0)$ lie on the graph of g, the points $(-4, 0)$ and $(0, 2)$ lie on the graph of g^{-1}. Notice that the point $(4, 4)$ lies on both lines. The graphs are shown below.

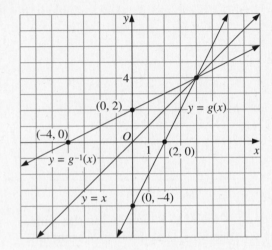

1. Give the coordinates of two more points on the graph of $y = g(x)$, and give the coordinates of their reflections on the graph of $y = g^{-1}(x)$.

2. Write the function $g^{-1}(x)$ in point-slope form.

Graph the function and its inverse in the same coordinate plane.

3. $f(x) = \dfrac{2}{3}x$ **4.** $g(x) = 9 - x$ **5.** $h(x) = -\dfrac{4}{5}x + 20$

Finding the equation of an inverse function

Given the equation for a function f, you can find the equation for f^{-1} by replacing $f(x)$ with y and solving the resulting equation for x.

Example 2

Let $f(x) = 24 + 3x$. Find the equation for f^{-1}.

▪ Solution ▪

Step 1: Rewrite the given function using y instead of $f(x)$.

$$y = 24 + 3x$$

Step 2: Solve for x.

$$y = 24 + 3x$$
$$y - 24 = 3x$$
$$\frac{1}{3}(y - 24) = \frac{1}{3}(3x)$$
$$\frac{1}{3}y - 8 = x$$

Step 3: Switch x and y so that x is the independent variable of the inverse function.

$$y = \frac{1}{3}x - 8$$

Step 4: Replace y with $f^{-1}(x)$.

$$f^{-1}(x) = \frac{1}{3}x - 8$$

An equation for $f^{-1}(x)$ is $f^{-1}(x) = \frac{1}{3}x - 8$.

6. Graph the inverse functions f and f^{-1} found in Example 2 in the same coordinate plane. Are the graphs reflections of each other over the line $y = x$? Where do the graphs intersect?

7. The function $g(x) = 2x - 4$ was used in Example 1. Use the method of Example 2 to find an equation for $g^{-1}(x)$.

8. Find an equation for the inverse of the function given in Exercise 3.

9. Find an equation for the inverse of the function given in Exercise 4.

10. Find an equation for the inverse of the function given in Exercise 5.

A recipe for cooked rice calls for 2 cups of water for every cup of rice. Use this information for Exercises 11–13.

11. Find an equation that gives the amount of water, w, in terms of the amount of rice, r. Identify the dependent and independent variables.

12. Solve the equation you wrote in Exercise 11 for r. The resulting equation gives the amount of __?__ in terms of the amount of __?__ . Identify the independent and dependent variables in this equation.

13. **Writing** Explain why the functions in Exercises 11 and 12 are inverses of each other.

14. **Mathematics Journal** What is the relationship between the slope of a linear function and the slope of its inverse? Use a sketch to illustrate your answer.

Spiral Review

TECHNOLOGY The table at the right shows the amount of electricity, in billions of kilowatt-hours, generated in the United States by nuclear power plants.
(Section 3.5)

x = years since 1980	y = billions of kilowatt-hours
0	255
5	384
10	577
11	613
12	619

15. Enter the data pairs (x, y) into a graphing calculator or graphing software. Find an exponential function that models the data.

16. Predict the amount of electricity that will be produced in the United States by nuclear power plants in the year 2000.

17. Estimate the amount of electricity that was produced in the United States by nuclear power plants in the year 1976.

For Exercises 18–20, find the point-slope equation of the line passing through the given points. *(Section 2.3)*

18. (3, 6) and (2, 8)

19. (–5, 2) and (–3, –4)

20. (4, –10) and (0, 6)

21. Suppose you deposit $650 in a savings account earning 5% annual interest compounded continuously. Find the amount of money in your account after 8 years, assuming you make no more deposits or withdrawls. *(Section 3.4)*

Section 4.2

Using Inverses of Exponential Functions

Learn how to . . .

- find inverses of exponential functions

- evaluate logarithms

So you can . . .

- understand the relationship between exponential functions and logarithmic functions

Application

Chemists use the pH scale to measure the acidity or alkalinity of a substance. The pH scale is logarithmic.

Terms to Know

Example / Illustration

Logarithmic function (p. 149) the inverse of an exponential function	
Base-b logarithm (p. 149) the inverse of the exponential function $f(x) = b^x$ (It is written $\log_b x$, where $b > 0$ and $b \neq 1$.)	The inverse of the function $y = b^x$ is the function $y = \log_b x$.
Common logarithm (p. 150) the base-10 logarithm of a positive number N (It is denoted $\log N$, rather than $\log_{10} N$.)	$10^2 = 100$, so $\log 100 = 2$
Natural logarithm (p. 150) the base-e logarithm of a positive number N (It is denoted $\ln N$, rather than $\log_e N$.)	$e^{0.5} \approx 1.65$, so $\ln 1.65 \approx 0.5$

Study Guide, ALGEBRA 2: EXPLORATIONS AND APPLICATIONS

Exponential and logarithmic form

For positive numbers a and b, with $b \neq 1$, the following two equations are equivalent.

exponential form: $b^a = x$ (for example, $3^2 = 9$)

logarithmic form: $\log_b x = a$ (for example, $\log_3 9 = 2$)

In other words, $\log_b x = a$ if and only if $b^a = x$

with annotations: *base*, *exponent*, *power of b*

Example 1

Write each equation in logarithmic form.

a. $6^{-3} = \dfrac{1}{216}$ **b.** $2^x = 8$

■ Solution ■

a. Use the equivalence $\log_b x = a$ if and only if $b^a = x$. Note that $b = 6$, $a = -3$, and $x = \dfrac{1}{216}$ in this case.

The logarithmic form of the exponential equation $6^{-3} = \dfrac{1}{216}$ is $\log_6 \dfrac{1}{216} = -3$.

b. The logarithmic form of the exponential equation $2^x = 8$ is $\log_2 8 = x$.

---

1. Which of the following is the logarithmic form of the exponential equation $9^{3/2} = 27$?

A. $\log_{27} 9 = \dfrac{3}{2}$ **B.** $\log_9 27 = \dfrac{3}{2}$ **C.** $\log_{3/2} 9 = 27$ **D.** $\log_9 \dfrac{3}{2} = 27$

2. Which of the following is the exponential form of the logarithmic equation $\log_8 0.5 = -\dfrac{1}{3}$?

A. $8^{-1/3} = 0.5$ **B.** $0.5^{-1/3} = 8$ **C.** $\left(-\dfrac{1}{3}\right)^8 = 0.5$ **D.** $8^{0.5} = -\dfrac{1}{3}$

Write each equation in logarithmic form.

3. $625^{0.25} = 5$ **4.** $19^0 = 1$ **5.** $(1.2)^2 = 1.44$ **6.** $\left(\dfrac{3}{5}\right)^{-3} = \dfrac{125}{27}$

Write each equation in exponential form.

7. $\log_2 32 = 5$ **8.** $\log_{0.5} 8 = -3$ **9.** $\log_5 5 = 1$ **10.** $\log_{0.25} 256 = -4$

Evaluating logarithms

You can evaluate a logarithmic expression by first writing it as an equation and then rewriting this logarithmic equation in exponential form.

Example 2

Evaluate $\log_{25} 0.2$.

■ Solution ■

Step 1: Let $\log_{25} 0.2 = x$.

Step 2: Write the logarithmic equation in exponential form: $25^x = 0.2$.

Step 3: Rewrite both sides of the exponential equation using the same base.

$$25^x = 0.2 \qquad \leftarrow 25 = 5^2 \text{ and } 0.2 = \frac{1}{5}$$

$$(5^2)^x = \frac{1}{5} \qquad \leftarrow \frac{1}{5} = 5^{-1}$$

$$5^{2x} = 5^{-1} \qquad \leftarrow (a^m)^n = a^{mn}$$

Step 4: Find the value of x that makes the two exponents the same. *(Note: $b^a = b^c$ if and only if $a = c$.)*

$$2x = -1$$

$$x = -\frac{1}{2}$$

Therefore, $\log_{25} 0.2 = -\frac{1}{2}$.

Note: You can verify the solution in Example 2 by using a scientific calculator to show that $25^{-1/2} = 0.2$.

Evaluate each logarithm.

11. $\log_{12} 144$

12. $\log_9 \dfrac{1}{27}$

13. $\log 1000$

14. $\log_{2/3} \dfrac{9}{4}$

15. $\ln e^{-7}$

16. $\log_{0.5} 128$

TECHNOLOGY Use a calculator to find the value of each logarithm to the nearest hundredth.

17. $\log 47$

18. $\log 0.0346$

19. $\ln 246$

20. $\ln \dfrac{7}{9}$

For each function:
a. Graph the function and its inverse on the same coordinate plane.
b. Find an equation for the inverse.

21. $y = 3^x$

22. $y = \left(\dfrac{1}{5}\right)^x$

For each function:
a. Graph the function and its inverse on the same coordinate plane.
b. Find an equation for the inverse. *(Section 4.1)*

23. $f(x) = 2x$ **24.** $g(x) = 0.8x$ **25.** $h(x) = x - 12$ **26.** $y = -\dfrac{2}{3}(x - 6)$

Let $A = \begin{bmatrix} 4 & 0 & 1 \\ 2 & 2 & 0 \\ 1 & 6 & 0 \end{bmatrix}$ and $B = \begin{bmatrix} -2 & 0 & 5 \\ 3 & 1 & -1 \\ 0 & 4 & 3 \end{bmatrix}$. **Find each matrix.** *(Sections 1.3 and 1.4)*

27. $A + B$ **28.** $B - A$ **29.** $3A + B$ **30.** AB

Tell whether each equation represents growth that is *linear*, *exponential*, or *neither*. *(Section 3.1)*

31. $y = 3(5^3)x$ **32.** $y = 3x^5$ **33.** $y = 5(2^x)$ **34.** $y = 2 + 5x$

Working with Logarithms

Learn how to . . .

- use the properties of logarithms

So you can . . .

- understand logarithmic scales

Application

The Mohs scale measures the hardness of rocks. The scale is logarithmic. Each number represents an increase in hardness by a factor of 10.

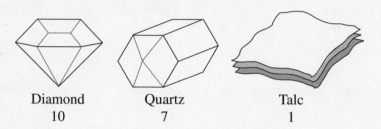

Diamond	Quartz	Talc
10	7	1

UNDERSTANDING THE MAIN IDEAS

Properties of logarithms

Let M, N, and b be positive numbers with $b \neq 1$. Then:

Property:

$$\log_b MN = \log_b M + \log_b N$$

$$\log_b \frac{M}{N} = \log_b M - \log_b N$$

$$\log_b M^k = k \log_b M$$

Example:

$$\log_2 35 = \log_2 7 + \log_2 5$$

$$\log_8 \frac{2}{5} = \log_8 2 - \log_8 5$$

$$\log_6 5^3 = 3 \log_6 5$$

The first property shown above is called the *product property*. The property shows that the logarithm of a product is the *sum* of the logarithms of the factors. The second property is called the *quotient property*. It shows that the logarithm of a quotient is the *difference* of the logarithms of the dividend and the divisor. The third property is called the *power property*. It shows that the logarithm of a power is the product of the exponent and the logarithm of the base.

Example 1

Write $\log_n 18 - 2 \log_n 3 + 4 \log_n 2$ as a logarithm of a single number.

■ Solution ■

$$\log_n 18 - 2\log_n 3 + 4\log_n 2 = \log_n 18 - \log_n 3^2 + \log_n 2^4 \qquad \leftarrow \text{power property}$$
$$= \log_n 18 - \log_n 9 + \log_n 16$$
$$= \log_n \frac{18}{9} + \log_n 16 \qquad \leftarrow \text{quotient property}$$
$$= \log_n 2 + \log_n 16$$
$$= \log_n 2(16) \qquad \leftarrow \text{product property}$$
$$= \log_n 32$$

Write each expression in terms of $\log_9 p$, $\log_9 q$, and $\log_9 r$.

1. $\log_9 p^2$

2. $\log_9 q^{0.6}r^{0.4}$

3. $\log_9 \dfrac{r^3}{q^6}$

4. $\log_9 \dfrac{q^4 r^2}{p^{1/3}}$

5. $\log_9 \dfrac{1}{p}$

6. $\log_9 p^5 q^{-3}$

For Exercises 7–12, write as a logarithm of a single number or expression.

7. $\log_b 8 - \log_b 4$

8. $\log_b \dfrac{2}{3} + \log_b 150$

9. $-\dfrac{1}{3}\log_b 125$

10. $9 \ln x^3 + 10 \ln x^{0.3}$

11. $3 \ln 5 - \dfrac{3}{2}\ln 25$

12. $3 \ln 6 - 2 \ln 3$

13. Technology Graph $y = \log 2x + \log 3x$ and $y = \log 5x$ using a graphing calculator or graphing software. Are the graphs of the two functions the same? If not, why not?

For a certain number b, $\log_b 5 = 0.774$, $\log_b 2 = 0.333$, and $\log_b 3 = 0.528$. Find each of the following.

14. $\log_b 24$

15. $\log_b 27$

16. $\log_b \dfrac{1}{5}$

17. $\log_b 2.5$

Applications involving logarithms

In a variety of real-world situations, the relationship between measured quantities is given by an equation or formula involving logarithms.

Example 2

The loudness of sound can be measured in "bels." A sound that is barely perceptible to the human ear has a loudness measure of 0 bels. The bel scale increases by powers of 10, so that an increase of 1 unit on the bel scale corresponds to a *tenfold* increase in the intensity of the sound waves. The bel scale relates loudness L and intensity I using the equation

$$L = \log \frac{I}{I_0}.$$

where I_0 is the intensity of the softest sound a human can hear. If the loudness of normal conversation is 6 bels, how many times greater is the intensity of amplified music measuring 13 bels?

Solution

Let I_C and I_M represent the intensity of the conversation and the intensity of the music, respectively. We need to find the ratio $\frac{I_M}{I_C}$.

Step 1: Find $\log \frac{I_M}{I_C}$.

By the quotient property of logarithms, $\log \frac{I_M}{I_C} = \log I_M - \log I_C$.

Using the formula, $6 = \log \frac{I_C}{I_0}$ and $13 = \log \frac{I_M}{I_0}$.

By the quotient property, $6 = \log I_C - \log I_0$ or $\log I_C = 6 + \log I_0$.
Similarly, $\log I_M = 13 + \log I_0$.

Now substitute these results into the equation for $\log \frac{I_M}{I_C}$.

$$\log \frac{I_M}{I_C} = \log I_M - \log I_C$$
$$= (13 + \log I_0) - (6 + \log I_0)$$
$$= 13 + \log I_0 - 6 - \log I_0$$
$$= 7$$

Step 2: Rewrite the result of Step 1 in exponential form.

$$\log \frac{I_M}{I_C} = 7 \quad \rightarrow \quad \frac{I_M}{I_C} = 10^7$$

The intensity of the sound waves from amplified music is about 10^7 (10 *million*) times the intensity of the sound waves from normal conversation.

18. The loudness of a race car is measured at 12.5 bels. The loudness of a car horn at a distance of 6 m is 11 bels. How many times more intense is the sound of the race car than the sound of the car horn?

19. You may have heard of the loudness measure *decibels* (dB). One decibel equals 10 bels. How many times more intense is the sound level in a gymnasium measured at 70 dB than the sound level in a library measured at 45 dB?

20. The bel rating of a soft whisper is 3. If the intensity of the sound is tripled, what is the new bel rating? Give your answer to the nearest hundredth.

....
Spiral Review

21. Write $2^3 = 8$ in logarithmic form. *(Section 4.2)*

22. Write $\log 0.0001 = -4$ in exponential form. *(Section 4.2)*

23. **Technology** Use a graphing calculator or graphing software to find the solution of $25e^{0.21x} = 71.4$ to the nearest hundredth. *(Section 3.4)*

Exponential and Logarithmic Equations

Application

To find the age of an object, archaeologists use carbon dating. This involves solving an exponential equation like $A = A_0\left(\dfrac{1}{2}\right)^{-5700t}$ for t.

Terms to Know	Example / Illustration
Extraneous solution (p. 165) a solution that does not satisfy the original equation	When solving the equation $\log_2(x + 3) + \log_2(x - 3) = 4$, the potential solutions $x = 5$ and $x = -5$ are obtained. The potential solution $x = -5$ is extraneous because when -5 is substituted for x in the equation, the terms $\log_2(x + 3)$ and $\log_2(x - 3)$ become $\log_2(-2)$ and $\log_2(-7)$, respectively, which are both undefined.

UNDERSTANDING THE MAIN IDEAS

Solving exponential equations

Just as you have used inverse operations to solve equations in the past, you use logarithms to solve exponential equations. Recall that the logarithmic and the exponential functions are inverse functions.

Example 1

Solve $5e^{12x} - 17 = 23$.

▪ Solution ▪

Since the variable x is part of the exponent, we will need to use a logarithm to "undo" the exponentiation.

$5e^{12x} - 17 = 23$ ← Add 17 to both sides.

$5e^{12x} = 40$ ← Divide both sides by 5.

$e^{12x} = 8$ ← Take the natural (base-e) logarithm of both sides.

$\ln e^{12x} = \ln 8$

$12x \ln e = \ln 8$ ← Use the power property.

$12x = \ln 8$ ← $\ln e = 1$ (This is why we used the natural logarithm.)

$x = \dfrac{\ln 8}{12}$ ← Use a calculator.

$x \approx 0.173$

To the nearest hundredth, the solution is 0.17.

1. Check the solution found in Example 1 by substituting 0.17 for x in the original equation.

2. **Technology** Graph $y = 5e^{12x}$ and $y = 40$ on the same set of axes. Where do the graphs intersect?

Solve each equation. Round your answer to the nearest hundredth.

3. $4^x = 15$

4. $25(1.01)^{x/12} = 100$

5. $3 + 7(0.65)^{x+3} = 15$

6. $8(5^x) = 9(2^x)$

All living organisms contain a small amount of a radioactive isotope of the element carbon–14 (C^{14}). When an organism dies, the amount of C^{14} decays according to the equation $A = A_0 2^{-t/5700}$, where A_0 is the initial amount of C^{14} and t is the number of years since death.

7. An archaeologist tests a sample taken from a piece of wood found at an archaeological site and finds that it contains 0.8 picograms of C^{14}. A sample taken from a living piece of wood the same size would contain 1 picogram of C^{14}. About how old is the piece of wood found at the site?

8. An archaeologist tests a piece of fossilized bone for C^{14} and finds that $A = \dfrac{1}{9} A_0$. About how old is the bone?

Solving logarithmic equations

To solve a logarithmic equation, you use the exponential function to "undo" the logarithm.

Example 2

Solve $-1 + 4 \log_{81} (x + 5) = 2$.

■ Solution ■

Step 1: Use algebra to get the logarithm alone on one side of the equation.

$$-1 + 4 \log_{81} (x + 5) = 2 \qquad \leftarrow \text{Add 1 to both sides.}$$
$$4 \log_{81} (x + 5) = 3 \qquad \leftarrow \text{Divide both sides by 4.}$$
$$\log_{81} (x + 5) = \frac{3}{4}$$

Step 2: Rewrite the equation in exponential form.

$$\log_{81} (x + 5) = \frac{3}{4} \qquad \rightarrow \qquad x + 5 = 81^{3/4}$$

Step 3: Solve the exponential equation for the variable.

$$x + 5 = 81^{3/4} \qquad \leftarrow 81^{3/4} = (81^{1/4})^3 = (\sqrt[4]{81})^3 = 3^3$$
$$x + 5 = 27$$
$$x = 22$$

Step 4: Check the solution in the original equation to see if it is extraneous. Recall that logarithms are only defined for positive numbers.

$$-1 + 4 \log_{81} (x + 5) = 2$$
$$-1 + 4 \log_{81} (22 + 5) \stackrel{?}{=} 2$$
$$-1 + 4 \log_{81} 27 \stackrel{?}{=} 2$$
$$-1 + 4\left(\frac{3}{4}\right) \stackrel{?}{=} 2$$
$$-1 + 3 \stackrel{?}{=} 2$$
$$2 = 2 \checkmark$$

The solution is 22.

For Exercises 9–12, solve each equation. Round your answers to the nearest hundredth.

9. $7 + \log_4 x = 10$

10. $\log_5 x - \log_5 (x + 2) = 1$

11. $\log_2 (3x + 10) + 1 = 0$

12. $\log (x + 1) = \log (x - 2) + 1$

13. Technology Solve the equation in Exercise 12 by graphing $y = \log (x + 1)$ and $y = \log (x - 2) + 1$ in the same coordinate plane using a graphing calculator or graphing software. Find the x-coordinate of their point of intersection. Give your answer to the nearest hundredth.

Using the change-of-base formula

You can convert a base-b logarithm into a base-c logarithm by using the formula

$$\log_b M = \frac{\log_c M}{\log_c b}$$

where M, b, and c are positive numbers with $b \neq 1$ and $c \neq 1$. When using the formula, people usually let $c = 10$ (common logarithm) or $c = e$ (natural logarithm) because most scientific calculators have a common logarithm key and a natural logarithm key. Using the change-of-base formula, you can evaluate any logarithm.

Example 3

Evaluate $\log_5 15$.

■ Solution ■

Use the change-of-base formula with $b = 5$ and $M = 15$.

Using common logarithms ($c = 10$):   *Using natural logarithms ($c = e$):*

$$\log_5 15 = \frac{\log 15}{\log 5} \quad \leftarrow \text{ Use a calculator. } \rightarrow \quad \log_5 15 = \frac{\ln 15}{\ln 5}$$

$$\approx \frac{1.176}{0.699} \qquad\qquad\qquad\qquad\qquad \approx \frac{2.708}{1.609}$$

$$\approx 1.68 \qquad\qquad\qquad\qquad\qquad\qquad \approx 1.68$$

For Exercises 14–16, evaluate each logarithm. Round your answers to the nearest hundredth.

14. $\log_2 50$ **15.** $\log_{0.3} 27$ **16.** $\log_{2/5} 4$

17. Technology Use a graphing calculator or graphing software to graph $y = \log_6 (x + 2)$. (*Hint*: Use the change-of-base formula to rewrite the equation using natural or common logarithms.)

....
Spiral Review

Write as a logarithm of a single number or expression. *(Section 4.3)*

18. $3 \log_b 8 + \frac{1}{3} \log_b 27$ **19.** $5 \log x + 2 \log y - 4 \log z$

Graph each function. *(Section 3.3)*

20. $y = \frac{1}{2}(4^x)$ **21.** $y = \frac{1}{2}(4^{-x})$ **22.** $y = -\frac{1}{2}(4^x)$

Using Logarithms to Model Data

Learn how to . . .

- use logarithms to find exponential and power functions that model data

So you can . . .

- make predictions about real-life quantities

Application

Logarithms are used to find mathematical models of planetary motion.

Terms to Know	**Example / Illustration**
Power function (p. 170) a function having the form $y = ax^b$ where a and b are constants	$y = \frac{1}{4}x^3 \qquad p = 0.0235n^{3.8}$

UNDERSTANDING THE MAIN IDEAS

Deciding if data fit an exponential model

You can use an *exponential function* to model a set of data points (x, y) if there is a *linear relationship* between x and log y. That is, if there is a linear model that fits the points $(x, \log y)$, then an exponential model can be found that fits the points (x, y). The regression line $\log y = ax + b$ gives the exponential model $y = 10^b(10^a)^x$.

Example 1

Decide if an exponential model fits the data in the table. If so, write an equation of the model giving P as a function of t.

U.S. per Capita Gross National Product (in current dollars)		
Year	Years since 1945 (t)	GNP (dollars) (P)
1955	10	2486
1965	20	3268
1975	30	7401
1985	40	16,766
1990	45	21,737

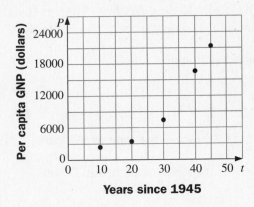

■ Solution ■

The scatter plot of the points (t, P) suggests an exponential model rather than a linear one. Make a table of data points $(t, \log P)$ and draw a scatter plot to see if a linear relationship exists for these points.

t	$\log P$
10	3.39
20	3.51
30	3.87
40	4.22
45	4.34

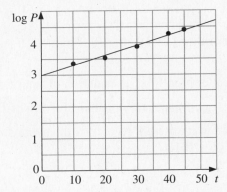

The points $(t, \log P)$ in the scatter plot appear to lie close to a line. An equation of the fitted line shown (as given by a graphing calculator, with correlation coefficient $r = 0.988$) is $\log P = 0.0289t + 3.03$.

Use the equation of the fitted line to write an equation that models the exponential relationship between t and P.

$\log P = 0.0289t + 3.03$
$\quad P = 10^{0.0289t + 3.03}$ ← Rewrite the equation in exponential form.
$\quad P = (10^{0.0289t})(10^{3.03})$ ← product rule of exponents: $b^{M+N} = b^M b^N$
$\quad P = (10^{3.03})(10^{0.0289})^t$ ← power rule of exponents: $b^{MN} = (b^M)^N$
$\quad P \approx 1070(1.07)^t$ ← Use a calculator.

An exponential equation modeling the per capita gross national product as a function of time is $P = 1070(1.07)^t$.

---

1. **Technology** Use a graphing calculator or graphing software to graph the points (P, t) given in Example 1 and the function $P = 1070(1.07)^t$. Do you think the function is a good model for the data?

2. The function $P = 1070(1.07)^t$ is an exponential growth model. What is the growth factor? In the years from 1955 to 1990, by what percent was the U. S. per capita gross national product growing per year?

Write y as a function of x.

3. $\log y = 3 + 0.5x$

4. $\log y = -x - 3$

5. $\ln y = 0.35 + 5.1x$

6. $\log y = -0.06x - 0.24$

For Exercises 7–10, use the data in the table.

Average Price of an Adult Ticket to the Movies		
Year	Years since 1940 (t)	Price (dollars) (P)
1948	8	0.36
1958	18	0.68
1967	27	1.22
1978	38	2.34
1988	48	4.11

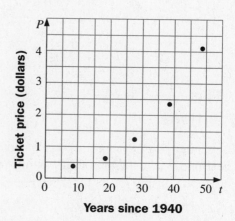

7. Make a scatter plot of the points $(t, \log P)$. What relationship exists between t and $\log P$?

8. Fit a line to your scatter plot. Find an equation of your fitted line.

9. Find an equation giving P as an exponential function of t.

10. Use your function from Exercise 9 to predict the average price of an adult movie ticket in the year 2010.

Deciding if data fits a power function

You can use a *power function* to model a set of data points (x, y) if there is a *linear relationship* between $\log x$ and $\log y$. That is, if there is a linear model that fits the points $(\log x, \log y)$, then a power model can be found that fits the points (x, y). The regression line $\log y = a \log x + b$ gives the power function $y = 10^b x^a$.

Study Guide, ALGEBRA 2: EXPLORATIONS AND APPLICATIONS

Example 2

Decide if a power model fits the data. If so, write an equation of the model giving y as a function of d.

Planet	d = mean distance from the sun (millions of miles)	y = time needed to circle the sun (Earth-years)
Mercury	36	0.24
Venus	67	0.62
Earth	93	1
Mars	142	1.88
Jupiter	484	11.9
Saturn	887	29.5
Uranus	1784	84.0
Neptune	2797	165

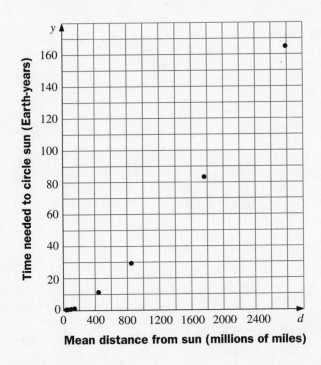

Mean distance from sun (millions of miles)

■ Solution ■

To see if a power curve is a suitable model, make a table of data points $(\log d, \log y)$ and draw a scatter plot to see if a linear relationship exists for these points.

(Solution continues on next page.)

■ **Solution** ■ *(continued)*

log *d*	log *y*
1.56	−0.62
1.83	−0.21
1.97	0
2.15	0.27
2.68	1.08
2.95	1.47
3.25	1.92
3.45	2.22

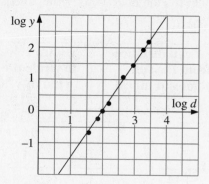

The points (log *d*, log *y*) in the scatter plot appear to have a strong linear relationship. This means that a power model fits the relationship between *d* and *y*. An equation of the fitted line shown (as given by a graphing calculator, with correlation coefficient *r* = 0.999) is log *y* = 1.50 log *d* − 2.96.

Use the equation of the fitted line to write an equation that models the power relationship between *d* and *y*.

$$\log y = 1.50 \log d - 2.96$$

$$\log y - 1.50 \log d = -2.96$$

$$\log \frac{y}{d^{1.50}} = -2.96 \qquad \leftarrow \text{quotient and power properties of logarithms}$$

$$\frac{y}{d^{1.50}} = 10^{-2.96} \qquad \leftarrow \text{Rewrite the equation in exponential form.}$$

$$y = (10^{-2.96})d^{1.50} \qquad \leftarrow \text{Use a calculator.}$$

$$y \approx 0.00110 d^{1.50}$$

An equation is $y = 0.00110 d^{1.50}$.

---

11. The mean distance from Pluto to the sun is 3666 million miles. How long (in Earth-years) is a year on Pluto?

Write *y* as a function of *x*.

12. $\log y = 3 - \log x$

13. $\log y = 0.5 \log x + 0.1$

14. $\log y = -0.4 + \frac{1}{3} \log x$

15. $\ln y = 2 + 3 \ln x$

The data in the table gives the speed, in miles per second, at which each of the first eight planets orbits the sun. Use this data for Exercises 16–19.

Planet	d = mean distance from the sun (millions of miles)	v = orbital velocity (in mi/s)
Mercury	36	29.75
Venus	67	21.76
Earth	93	18.51
Mars	142	14.99
Jupiter	484	8.12
Saturn	887	5.99
Uranus	1784	4.23
Neptune	2797	3.38

16. Make a scatter plot of the data pairs (log d, log v). What relationship exists between log d and log v?

17. Fit a line to your scatter plot. Find an equation of your fitted line.

18. Find an equation giving v as a power function of d.

19. The mean distance from Pluto to the sun is 3666 million miles. What is the orbital velocity of Pluto?

20. **Writing** To use the method in Example 2, both coordinates of all the data points must be positive. Explain why.

21. **Mathematics Journal** In Section 1.2, you learned a method for modeling exponential growth. Compare this method with the method of Example 1 in this section.

....
Spiral Review

For Exercises 22–25, evaluate each logarithm. Round your answers to the nearest hundredth. *(Section 4.4)*

22. $\log_2 19$

23. $\log_{20} 5$

24. $\log_4 0.28$

25. $\log_3 \frac{11}{12}$

26. **Writing** What information does the correlation coefficient give you about the least-squares line? Use examples in your explanation. *(Section 2.5)*

Evaluate each expression when $x = 4$ and $x = -3$. *(Toolbox, p. 780)*

27. $2x^2$

28. $-x^2$

29. $-0.5x^2$

30. $(2x)^2$

Chapter 4 Review ············

Complete these exercises for a review of Chapter 4. If you have difficulty with a particular problem, review the indicated section.

1. **Open-ended** Using any linear function of your choice, graph the function and its inverse in the same coordinate plane. *(Section 4.1)*

2. Find an equation for the inverse you graphed in Exercise 1. *(Section 4.1)*

3. The dollar cost C of the electricity used in a home in one month is given by $C = 7.2 + 0.114h$, where h is the number of kilowatt-hours used. Write the inverse of this function to give a formula for h in terms of C. *(Section 4.1)*

Write each equation in logarithmic form. *(Section 4.2)*

4. $\dfrac{1}{36} = 6^{-2}$

5. $e^3 \approx 20.1$

Write each equation in exponential form. *(Section 4.2)*

6. $\log_{25} 0.2 = -0.5$

7. $\log 1000 = 3$

For Exercises 8 and 9, evaluate each logarithm. *(Section 4.2)*

8. $\log_{16} 0.5$

9. $\ln \sqrt{e}$

10. **Writing** If $\ln x < \ln y$, what can you say about x and y? Explain. *(Section 4.3)*

11. Write $\log_b 15 - 2 \log_b 5 + \dfrac{3}{2} \log_b 9$ as a logarithm of a single number or expression. *(Section 4.3)*

12. If $\log_2 3 \approx 1.58$ and $\log_2 5 \approx 2.32$, estimate the value of $\log_2 30$ without using a calculator. *(Section 4.3)*

Solve each equation. Round your answers to the nearest hundredth. *(Section 4.4)*

13. $2 + 5^{3x} = 5.51$

14. $400(1.04)^{x/12} = 600$

15. $\log_9 (56 - x) - \log_9 x = 1.5$

16. $\log x + \log (x + 3) = 1$

Write y as a function of x. *(Section 4.5)*

17. $\log y = 0.7 + 0.3 \log x$

18. $\log y = 0.7 + 0.3x$

The table below shows the number of computers sold in the United States each year since 1990. *(Section 4.5)*

Number of years since 1990	Number of computers sold (millions)
0	3.90
1	4.88
2	5.85
3	6.75
4	8.23
5	9.53

19. Let t represent the number of years since 1990 and let n represent the number of computers sold (in millions). Make a scatter plot of the points $(t, \log n)$. What type of relationship exists between t and $\log n$?

20. Fit a line to the scatter plot. Find an equation of your fitted line.

21. Find an equation giving n as an exponential function of t.

22. Use your equation from Exercise 21 to predict the number of computers sold in the United States in 1996.

SPIRAL REVIEW   Chapters 1–4

You have a bowl of pebbles sitting at room temperature, 70°F. You put it in a freezer set at 0°F. As the pebbles cool, their Fahrenheit temperature decays exponentially with time. After 15 min in the freezer, the temperature of the pebbles is 50°F.

1. Write an exponential function in the form $C = ab^t$ modeling the Fahrenheit temperature after t minutes.

2. When did the temperature of the pebbles drop below 60°F?

3. When will the temperature of the pebbles reach 35°F?

Multiply.

4. $[6 \ -1 \ 3]\begin{bmatrix} 4 & 0 & 2 & 1 \\ 3 & 9 & 5 & 0 \\ 2 & -1 & 0 & 8 \end{bmatrix}$

5. $\begin{bmatrix} 2 & 0 & 3 \\ -2 & -4 & 0 \\ 0 & 1.5 & -1 \end{bmatrix}\begin{bmatrix} -4 & -4.5 & -12 \\ 2 & 2 & 6 \\ 3 & 3 & 8 \end{bmatrix}$

Match each description with one of the formulas below.

A. $A = Pe^{1.06t}$
B. $A = P(1.06)^t$
C. $A = P(1.005)^t$
D. $A = Pe^{0.06t}$
E. $A = P(1.005)^{12t}$

6. the value of an investment of P dollars after t years, if it earns interest at 6% compounded annually

7. the value of an investment of P dollars after t years, if it earns interest at 6% compounded monthly

8. the value of an investment of P dollars after t years, if it earns interest at 6% compounded continuously

Working with Simple Quadratic Functions

Learn how to . . .

- recognize and draw graphs of quadratic functions
- solve simple quadratic equations

So you can . . .

- examine the relationships between two quantities

Application

The surface area of a sphere with radius r is given by the formula S.A. $= 4\pi r^2$. Earth is a sphere with a radius of about 4000 mi. So the surface area of Earth is about $4\pi(4000)^2$, or about 200 million square miles.

Terms to Know	Example / Illustration
Quadratic function (p. 186) a function having the general form $y = ax^2$ (The graph of a quadratic function is a parabola.)	$y = 4\pi r^2$
Parabola (p. 186) the graph of a quadratic function	
Line of symmetry (p. 186) the line that can be drawn *through* a parabola such that the part of the parabola on one side of the line is a reflection of the part on the other side	The line of symmetry for the parabola shown above is the y-axis.
Vertex (p. 186) the point where the line of symmetry crosses a parabola	The vertex of the parabola shown above is the point $(0, 0)$.

Terms to Know	Example / Illustration
Square root of a number (p. 187) a value that when squared is the number	The square roots of 0.25 are 0.5 and –0.5 since $(0.5)^2 = 0.25$ and $(-0.5)^2 = 0.25$.

UNDERSTANDING THE MAIN IDEAS

A simple quadratic function has the form $y = ax^2$. The graph of $y = ax^2$ is a parabola that is symmetric about the y-axis. When graphing a function of the form $y = ax^2$, $a \neq 1$, it helps to compare it to the graph of $y = x^2$. For functions with $a > 0$, the graph opens upward like the graph of $y = x^2$. For functions with $a < 0$, the graph opens downward. If $|a| > 1$, the graph will be narrower than the graph of $y = x^2$. If $|a| < 1$, the graph will be wider than the graph of $y = x^2$.

Example 1

Match each graph with its equation.

A. $y = x^2$ **B.** $y = 4x^2$ **C.** $y = 0.25x^2$

a.
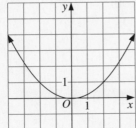

b.
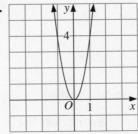

c.
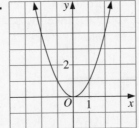

Solution

a. The graph is wider than the graph of $y = x^2$, indicating a vertical shrink, and it passes through the point (2, 1). The correct equation of the graph is $y = 0.25x^2$, choice C.

b. The graph is narrower than the graph of $y = x^2$, indicating a vertical stretch, and it passes through the point (1, 4). The correct equation of the graph is $y = 4x^2$, choice B.

c. This is the graph of the equation $y = x^2$, choice A.

Match each graph with its equation.

A. $y = 3x^2$ **B.** $y = \frac{1}{3}x^2$ **C.** $y = x^2$ **D.** $y = -3x^2$

1.
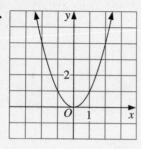

2.
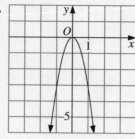

3.
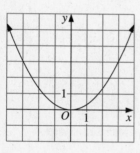

4.
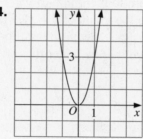

Solving a quadratic equation

When solving an equation of the form $ax^2 = b$, the first step is to divide both sides by a. Then you must take the square root of both sides of the resulting equation in order to obtain x by itself.

Example 2

Solve each equation. Give solutions to the nearest tenth when necessary.

a. $x^2 = 9$ **b.** $-3x^2 = -18$ **c.** $36x^2 = 4$

■ Solution ■

a. Since the coefficient of x^2 is already 1, solve the equation by taking the square root of both sides.

$$x^2 = 9$$

$$x = \pm\sqrt{9} \quad \leftarrow \quad \sqrt{x^2} = x; \text{ Both the positive and negative}$$
square root of 9 are solutions.

$$x = \pm 3$$

(Solution continues on next page.)

■ **Solution** ■ *(continued)*

b. $-3x^2 = -18$

$$\frac{-3x^2}{-3} = \frac{-18}{-3}$$

$$x^2 = 6$$

$$x = \pm\sqrt{6}$$

$$x \approx \pm 2.4$$

c. $36x^2 = 4$

$$\frac{36x^2}{36} = \frac{4}{36}$$

$$x^2 = \frac{1}{9}$$

$$x = \pm\sqrt{\frac{1}{9}}$$

$$x = \pm\frac{1}{3}$$

Solve each equation. Give solutions to the nearest tenth when necessary.

5. $x^2 = 144$

6. $-5x^2 = -125$

7. $15x^2 = 60$

8. $\frac{1}{4}x^2 - 9 = 0$

9. $-3x^2 = -27$

10. $4x^2 = 1$

11. $x^2 = 8$

12. $5x^2 = 15$

13. $2x^2 = 24$

Finding the equation of a parabola

If you know or can determine the ordered pair of at least one point that lies on a parabola, you can use the coordinates to write an equation of the graph.

Example 3

The graph at the right is that of a function of the form $y = ax^2$. Find the value of a.

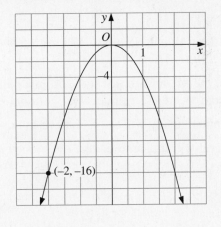

$(-2, -16)$

■ Solution ■

Since the point $(-2, -16)$ lies on the parabola, we know that $y = -16$ when $x = -2$.

$\quad y = ax^2 \qquad \leftarrow$ Substitute -16 for y and -2 for x.

$\quad -16 = a(-2)^2$

$\quad -16 = 4a$

$\quad -4 = a$

The equation of the parabola is $y = -4x^2$.

Each graph has an equation of the form $y = ax^2$. Find a.

14.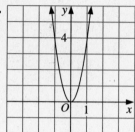

15.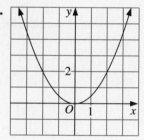

....
Spiral Review

16. Suppose $\log y + 1.2 = 3 \log x$. Write y as a function of x. *(Section 4.5)*

17. Graph each equation. *(Section 2.2)*

   a. $y = 2x + 3$ **b.** $y = 2x - 1$

18. How are the graphs in Exercise 17 alike? How are they different? *(Section 2.2)*

19. Supppose the Math Club invests \$250 in a bank account that offers 6% interest compounded continuously. *(Section 3.4)*

   a. How much money will the club have in the account after one year?

   b. How long will it take to double the money in the account?

Section 5.2

GOAL

Learn how to . . .

- graph equations in the form $y = a(x - h)^2 + k$

- solve quadratic equations

So you can . . .

- solve problems

Translating Parabolas

Application

A ball tossed straight up into the air will reach a maximum height and then fall to the ground. The relationship between the height of the ball and the time since it was tossed is quadratic, modeled by an equation such as $h = -16(t - 0.5)^2 + 8$.

Terms to Know

Example / Illustration

Terms to Know	Example / Illustration
Vertex form (p. 193) the form $y = a(x - h)^2 + k$ of a quadratic equation (The variables h and k are the coordinates of the vertex, (h, k).)	The equation $h = -16(t - 0.5)^2 + 8$ is the vertex form of the equation of a parabola with vertex at $(0.5, 8)$.
Minimum value (p. 195) the value k in the equation $y = a(x - h)^2 + k$ when $a > 0$ (The point (h, k) is the lowest point on the graph of $y = a(x - h)^2 + k$ when $a > 0$.)	The minimum value of the quadratic function $y = 2(x - 3)^2 + 4$ is 4. (The point $(3, 4)$ is the lowest point on the graph of the function.)
Maximum value (p. 195) the value k in the equation $y = a(x - h)^2 + k$ when $a < 0$ (The point (h, k) is the highest point on the graph of $y = a(x - h)^2 + k$ when $a < 0$.)	The maximum value of the quadratic function $y = -5(x - 3)^2 - 3$ is -3. (The point $(3, -3)$ is the highest point on the graph of the function.)

UNDERSTANDING THE MAIN IDEAS

The graph of the equation $y = a(x - h)^2 + k$ is a parabola with vertex at (h, k) and line of symmetry at $x = h$. The parabola opens upward if $a > 0$ and opens downward if $a < 0$.

Example 1

Match each quadratic equation with one of the graphs below.

a. $y = -3x^2 + 2$ **b.** $y = 3(x + 2)^2$ **c.** $y = (x - 2)^2 - 1$

A. 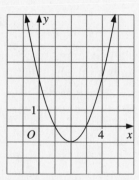   **B.** 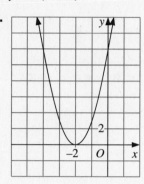   **C.**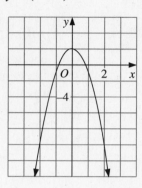

■ Solution ■

a. The graph of $y = -3x^2 + 2$ opens downward since $-3 < 0$. When written in vertex form, the equation is $y = -3(x - 0)^2 + 2$, so the vertex of the parabola is at $(0, 2)$. Therefore, the correct graph is choice C.

b. The graph of $y = 3(x + 2)^2$ is a parabola that opens upward since $3 > 0$. When written in vertex form, the equation is $y = 3(x + 2)^2 + 0$, so the vertex is at $(-2, 0)$. The correct graph is choice B.

c. The graph of $y = (x - 2)^2 - 1$ is a parabola that opens upward, with vertex at $(2, -1)$. The correct graph is choice A.

Match each quadratic equation with one of the graphs below.

1. $y = 3(x + 2)^2 - 2$ **2.** $y = 3x^2 + 2$ **3.** $y = 3(x - 2)^2 + 2$

A. 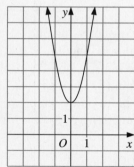   **B.** 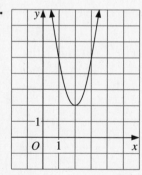   **C.**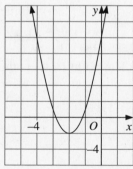

Minimum and maximum values

For a quadratic function whose graph opens upward, the minimum value of the function is the y-coordinate of the vertex of the parabola. That is, for a quadratic function written in vertex form in which $a > 0$, the *minimum* value of the function is k. Similarly, for a quadratic function written in vertex form in which $a < 0$, the *maximum* value of the function is k.

Example 2

What is the maximum value or minimum value of each function?

a. $f(x) = -0.5(x + 1)^2 + 5$ **b.** $f(x) = 2x^2 - 6$ **c.** $f(x) = -2(x - 3)^2 + 14$

■ Solution ■

a. Because the vertex (h, k) represents the highest point on the graph of $y = a(x - h)^2 + k$ when $a < 0$, the vertex of the parabola is $(-1, 5)$ and the *maximum* value of the function is 5.

b. The vertex of the parabola is $(0, -6)$ and $a = 2$. Therefore, the *minimum* value of the function is -6.

c. The vertex of the parabola is $(3, 14)$ and $a = -2$. Therefore, the *maximum* value of the function is 14.

What is the maximum value or minimum value of each function?

4. $f(x) = -2(x + 1)^2 + 3$ **5.** $f(x) = 0.5(x - 1)^2 - 2$ **6.** $f(x) = 0.2(x - 3)^2 - 2$

7. $f(x) = -3(x + 4)^2 + 1$ **8.** $f(x) = -0.5(x + 2)^2 + 4$ **9.** $f(x) = 3(x + 3)^2 + 4$

Example 3

Describe the graph of each function. Make a sketch of the graph.

a. $y = -0.5(x + 1)^2 + 5$ **b.** $y = 2(x - 6)^2$

■ Solution ■

a. Rewrite the equation as $y = -0.5(x - (-1))^2 + 5$. From the vertex form of the function, it can be seen that the parabola opens down, the vertex is $(-1, 5)$, and the line of symmetry is $x = -1$. Now find several points on the parabola; use the fact that the parabola is symmetric over the line of symmetry. When $x = -2$ and $x = 0$, the corresponding y-value is $y = -0.5(\pm 1)^2 + 5$, or 4.5. When $x = -3$ and $x = 1$, the corresponding y-value is $y = -0.5(\pm 2)^2 + 5$, or 3. So the points $(-3, 3)$, $(-2, 4.5)$, $(0, 4.5)$, and $(1, 3)$ are on the parabola. The graph is shown at the right.

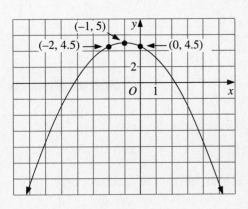

(Solution continues on next page.)

■ **Solution** ■ *(continued)*

b. Rewrite the equation as
$y = 2(x - 6)^2 + 0$. The parabola opens
up, the vertex is (6, 0), and the line of
symmetry is $x = 6$. When $x = 5$ or
$x = 7$, the corresponding y-value is
$y = 2(\pm1)^2 = 2$. When $x = 4$ or $x = 8$,
the corresponding y-value is
$y = 2(\pm2)^2 = 8$. So the points (4, 8),
(5, 2), (7, 2), and (8, 8) are on the
parabola. The graph is shown at
the right.

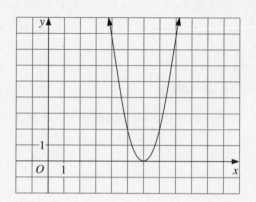

Describe the graph of each function. Make a sketch of each graph.

10. $y = 3(x - 2)^2 + 1$

11. $y = -2(x + 2)^2 - 3$

12. $y = \dfrac{1}{5}(x - 2)^2 + 5$

13. $y = -(x - 4)^2 - 5$

For Exercises 14 and 15, write an equation in the form $y = a(x - h)^2 + k$ for each parabola shown.

14.

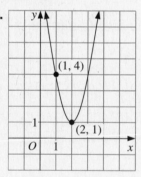

15.

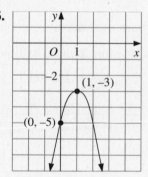

·····················
Spiral Review

Solve each equation. *(Section 5.1)*

16. $2x^2 = 18$

17. $3x^2 - 8 = 1$

18. $x^2 = 0.64$

Rewrite each expression in factored form. *(Toolbox, page 782)*

19. $25x + 5$

20. $7x^2 + 14x$

21. $4xy + 10x^2y$

Find the inverse of each function. *(Sections 4.1 and 4.2)*

22. $f(x) = 2x + 3$

23. $g(x) = 10^x$

24. $h(x) = e^{2x}$

Quadratic Functions in Intercept Form

Learn how to . . .

- write quadratic
  equations in
  intercept form

- maximize or
  minimize quadratic
  functions

So you can . . .

- find the maximum
  or minimum value
  for certain
  problems

Application

When the price of a ticket to a school dance is $3.00, the number of students expected to attend is 200. Each price increase of $.50 lowers the number of students who attend by 20. An equation for the expected revenue from the dance (ticket price × number of students) is $R = (3 + 0.5x)(200 - 20x)$, where x = the number of $.50 increases.

Terms to Know	Example / Illustration

x-intercept (p. 199)
    the x-coordinate of a point where a graph crosses
    the x-axis
	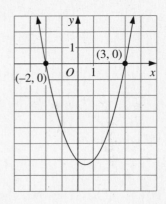
	
	The x-intercepts of the graph above are −2 and 3.

Intercept form (p. 200)
    the form $y = a(x - p)(x - q)$ of a quadratic function
    (This form of a quadratic function allows you to read
    the x-intercepts, p and q, directly.)
	$y = 2(x + 3)(x - 5)$

UNDERSTANDING THE MAIN IDEAS

You can use the intercept form of a quadratic function to graph a parabola by finding the x-intercepts and the vertex. The x-coordinate of the vertex is $x = \frac{p+q}{2}$.

The y-coordinate of the vertex is found by substituting the value of $\frac{p+q}{2}$ for x in the quadratic function.

Example 1

Graph $y = (2x + 6)(x - 1)$.

■ Solution ■

Step 1: Rewrite $y = (2x + 6)(x - 1)$ in the form $y = a(x - p)(x - q)$.

$$y = (2x + 6)(x - 1)$$
$$= 2(x + 3)(x - 1)$$
$$= 2(x - (-3))(x - 1)$$

Step 2: Identify the x-intercepts and the coordinates of the vertex.

The x-intercepts are -3 and 1.

The x-coordinate of the vertex is:

$$x = \frac{-3 + 1}{2} = -1$$

The y-coordinate of the vertex is:

$$y = (2x + 6)(x - 1) \qquad \leftarrow \text{Substitute } -1 \text{ for } x.$$
$$= (2(-1) + 6)(-1 - 1)$$
$$= 4(-2)$$
$$= -8$$

The vertex is the point $(-1, -8)$.

Step 3: Graph the function $y = 2(x - (-3))(x - 1)$ as a parabola with the x-intercepts and vertex found in Step 2.

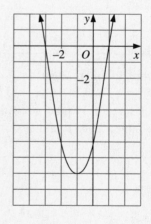

For each equation:

a. Find the *x*-intercept(s).

b. Find the vertex.

c. Sketch the graph.

1. $y = 3(x - 4)(x + 1)$ **2.** $y = -2(x - 1)(x + 3)$

3. $y = -(x - 2)(x + 2)$ **4.** $y = (2x + 2)(3x + 6)$

5. $y = (4 - x)(2x + 8)$ **6.** $y = 0.5(x - 4)(x - 4)$

Writing an equation from a graph

To write the equation for the graph of a parabola, first identify the *x*-intercepts of the graph. These are the values of *p* and *q* in the intercept form $y = a(x - p)(x - q)$. Then identify the vertex of the graph and substitute the coordinates for *x* and *y*, and then solve for *a*.

Example 2

Write an equation for the parabola shown below.

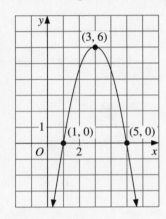

■ Solution ■

Use the intercept form $y = a(x - p)(x - q)$.

Step 1: Identify the *x*-intercepts and substitute them for *p* and *q* in the intercept form.

From the graph, the *x*-intercepts are 1 and 5. Substitute 1 for *p* and 5 for *q* in the intercept form.

$$y = a(x - 1)(x - 5)$$

(Solution continues on next page.)

■ **Solution** ■ *(continued)*

Step 2: Identify the vertex and use its coordinates to find the value of *a* in the equation from Step 1.

The vertex is (3, 6), so let $x = 3$ and $y = 6$.

$$6 = a(3 - 1)(3 - 5)$$
$$6 = a(2)(-2)$$
$$6 = -4a$$
$$a = -1.5$$

An equation in intercept form for the parabola is $y = -1.5(x - 1)(x - 5)$.

Write an equation for each graph.

7.
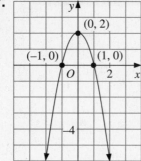

8.
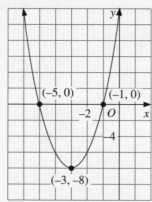

9.
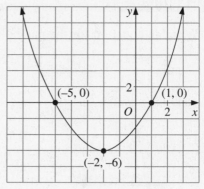

Example 3

When the price of a ticket to a school dance is $3.00, the number of students expected to attend is 200. For each price increase of $.50, 20 fewer students are expected to attend. An equation modeling the expected revenue *R* from the dance (ticket price times number of students) is found to be

$$R = (3 + 0.5x)(200 - 20x)$$

where *x* = number of $.50 increases above $3.00 in the ticket price. What ticket price will give the maximum revenue and what will be the revenue at this price?

■ Solution ■

The equation $R = (3 + 0.5x)(200 - 20x)$ is quadratic. The number of $.50 price increases that will maximize the revenue is the x-coordinate of the vertex of the graph $R = (3 + 0.5x)(200 - 20x)$.

Step 1: Find the x-coordinate of the vertex.

The x-coordinate of the vertex is the average of the x-intercepts of the graph. Find the x-intercepts by setting the factors $3 + 0.5x$ and $200 - 20x$ from the revenue equation equal to 0.

$$3 + 0.5x = 0 \qquad 200 - 20x = 0$$
$$0.5x = -3 \qquad\qquad -20x = -200$$
$$x = -6 \qquad\qquad\quad x = 10$$

So the x-coordinate of the vertex is $x = \dfrac{-6 + 10}{2}$, or 2.

Step 2: Use the x-coordinate found in Step 1 to find the ticket price and the maximum revenue.

ticket price: $3 + 0.5(2) = 4$

$$R = (3 + 0.5(2))(200 - 20(2))$$
$$= (3 + 1)(200 - 40)$$
$$= 4(160)$$
$$= 640$$

A ticket price of $4.00 results in a maximum revenue of $640.

---

10. Use the revenue equation $R = (3 + 0.5x)(200 - 20x)$ to find the revenue that results from a ticket price of $3.00 (*no* increase in ticket price).

....
Spiral Review

Graph each function in the same coordinate plane. Describe the relationships among the four graphs. *(Section 5.2)*

11. $y = x^2$ **12.** $y = (x + 1)^2$

13. $y = (x + 1)^2 + 2$ **14.** $y = -(x + 1)^2 + 2$

Solve each equation. Give solutions to the nearest tenth when necessary. *(Sections 5.1 and 5.2)*

15. $9x^2 = 144$ **16.** $5(x - 1)^2 = 80$ **17.** $12(x + 0.5)^2 - 36 = 0$

Write as a logarithim of a single number or expression. *(Section 4.3)*

18. $-\dfrac{1}{2} \log_b 16$ **19.** $\log_b m - 3 \log_b n$ **20.** $2 \log_b x + 3 \log_b x$

Completing the Square

Learn how to . . .

- complete the square to write quadratic functions in vertex form

So you can . . .

- find maximums or minimums

Application

The height h (in feet) above the ground at time t (in seconds) of a ball thrown straight up into the air is given by the function

$$h = -16t^2 + 24t + 4.$$

A process called *completing the square* can be used to rewrite this function so that the maximum height of the ball can be easily identified.

Terms to Know	Example / Illustration
Standard form (p. 206) the form $y = ax^2 + bx + c$ of a quadratic function	$y = -2x^2 + 12x - 14$

UNDERSTANDING THE MAIN IDEAS

You can find the vertex of a parabola whose equation is given in the standard form $y = ax^2 + bx + c$ by using a process called *completing the square* to change the equation into the vertex form $y = a(x - h)^2 + k$.

Example 1

Write each function in vertex form.

  a. $y = 2x^2 + 8x + 3$ **b.** $y = -2x^2 - 6x + 5$

▪ Solution ▪

a. *Step 1:* Working with just the x^2-term and the x-term of the function, factor out the coefficient of the x^2-term.

$$y = 2x^2 + 8x + 3 \quad \rightarrow \quad y = 2(x^2 + 4x) + 3$$

(Solution continues on next page.)

■ **Solution** ■ *(continued)*

> **Step 2:** Complete the square for the expression inside the parentheses. (*Note:* Whatever number is added to complete the square must also be subtracted so as not to change the equation.)

$$y = 2(x^2 + 4x) + 3$$

$$y = 2(x^2 + 4x + \underline{\,?\,}) + 3 - \underline{\,?\,} \quad \leftarrow \quad \text{The number we need is the square of half the coefficient of the } x\text{-term.}$$

$$y = 2(x^2 + 4x + 4) + 3 - 8 \quad \leftarrow \quad (4 \div 2)^2 = 2^2 = 4; \text{ we have added } 2(4) = 8 \text{ so we must subtract 8.}$$

$$y = 2(x + 2)^2 - 5 \quad \leftarrow \quad x^2 + 4x + 4 = (x + 2)^2$$

The vertex form of the function is $y = 2(x + 2)^2 - 5$.

b. $y = -2x^2 - 6x + 5$

$$y = -2(x^2 + 3x) + 5$$

$$y = -2\left(x^2 + 3x + \frac{9}{4}\right) + 5 - (-2)\left(\frac{9}{4}\right) \quad \leftarrow \quad \left(\frac{3}{2}\right)^2 = \frac{9}{4}$$

$$y = -2\left(x^2 + 3x + \frac{9}{4}\right) + 5 + \frac{9}{2}$$

$$y = -2\left(x + \frac{3}{2}\right)^2 + \frac{19}{2}$$

The vertex form of the function is $y = -2\left(x + \frac{3}{2}\right)^2 + \frac{19}{2}$.

Write each function in vertex form.

1. $y = x^2 + 2x + 1$ **2.** $y = x^2 + 8x + 13$ **3.** $y = x^2 + 6x + 6$

4. $y = -2x^2 + 4x + 1$ **5.** $y = 5x^2 - 40x + 78$ **6.** $y = -5x^2 + 40x - 82$

Recall that rewriting a quadratic function in vertex form is useful when identifying the graph of the function.

Example 2

Match each equation with its graph. Explain your choices.

a. $y = 2x^2 - 12x + 9$ **b.** $y = -2x^2 + 12x - 9$ **c.** $y = 2x^2 + 12x + 9$

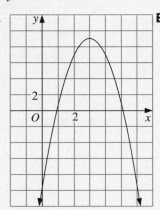 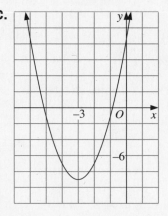

Solution

a. Rewrite the equation $y = 2x^2 - 12x + 9$ in vertex form. Then identify the vertex and use it to help identify the correct graph.

$y = 2x^2 - 12x + 9$ ← Complete the square.
$y = 2(x^2 - 6x) + 9$
$y = 2(x^2 - 6x + \underline{\;?\;}) + 9 - \underline{\;?\;}$

$y = 2(x^2 - 6x + 9) + 9 - 2(9)$ ← $\left(-\dfrac{6}{2}\right)^2 = (-3)^2 = 9$; we have added $2(9)$.

$y = 2(x - 3)^2 - 9$ ← vertex form

The vertex is $(3, -9)$. Notice that $a > 0$, so the parabola opens up. Therefore, the correct graph is choice B.

b. $y = -2x^2 + 12x - 9$
$y = -2(x^2 - 6x + \underline{\;?\;}) - 9 - \underline{\;?\;}$
$y = -2(x^2 - 6x + 9) - 9 - (-2(9))$ ← We must add and subtract $-2(9)$.
$y = -2(x - 3)^2 + 9$

The vertex is $(3, 9)$. Since $a < 0$, the parabola opens down. Therefore, the correct graph is choice A.

c. $y = 2x^2 + 12x + 9$
$y = 2(x^2 + 6x + \underline{\;?\;}) + 9 - \underline{\;?\;}$
$y = 2(x^2 + 6x + 9) + 9 - 2(9)$
$y = 2(x + 3)^2 - 9$

The vertex is $(-3, -9)$. Since $a > 0$, the parabola opens up. Therefore, the correct graph is choice C.

Match each equation with its graph. Explain your choices.

7. $y = x^2 - 4x + 3$ **8.** $y = -x^2 - 4x - 5$ **9.** $y = 2x^2 - 4x - 3$

A. 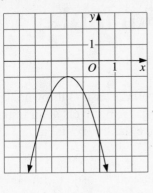   **B.** 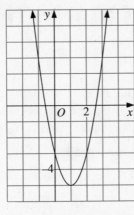   **C.**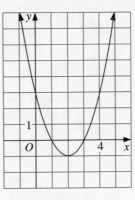

For Exercises 10 and 11, refer to the situation described in the Application.

10. Find the time in seconds it takes for the ball to reach its maximum height.

11. Find the maximum height that the ball reaches.

....
Spiral Review

Write an equation in intercept form for each parabola shown. *(Section 5.3)*

12. 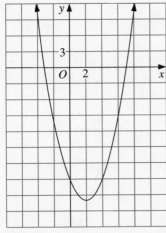   **13.**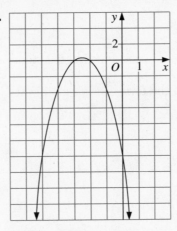

Solve each equation. Give solutions to the nearest tenth when necessary. *(Section 5.1)*

14. $9x^2 = 4$ **15.** $300 = 3x^2$ **16.** $\dfrac{2}{5}x^2 = \dfrac{5}{8}$ **17.** $2x^2 = 60$

Find the inverse of each function. *(Section 4.1)*

18. $f(x) = 3x + 1$ **19.** $f(x) = 3.2x - 1.4$ **20.** $f(x) = 2 - 2x$

Using the Quadratic Formula

Learn how to . . .

- solve equations using the quadratic formula

- use the discriminant to determine how many solutions an equation has

So you can . . .

- find the solutions of a quadratic equation

Application

A system of equations that models the path of a javelin after it is thrown describes the horizontal distance d (in feet) that the javelin travels at an initial speed of 100 ft/s and at an angle a (in degrees) above the horizontal as

$$d = -0.251a^2 + 16.5a + 53.4.$$

When the value of d is known, the quadratic formula can be used to find the corresponding value(s) of a, if any exist.

Terms to Know	**Example / Illustration**
Quadratic formula (p. 216) the formula $x = \dfrac{-b \pm \sqrt{b^2 - 4ac}}{2a}$ used to find the solutions of any quadratic equation of the form $ax^2 + bx + c = 0$	If $2x^2 + 9x + 1 = 0$, then $x = \dfrac{-9 \pm \sqrt{9^2 - 4(2)(1)}}{2(2)}$ $= \dfrac{-9 \pm \sqrt{81 - 8}}{4}$ $= \dfrac{-9 \pm \sqrt{73}}{4}$ Using a calculator, $x \approx -4.4$ and $x \approx -0.1$.
Discriminant (p. 217) the value $b^2 - 4ac$ in the quadratic formula (The sign of the discriminant can be used to determine the number of solutions of a quadratic equation.)	For the quadratic equation $2x^2 + 9x + 1 = 0$, $b^2 - 4ac = (9)^2 - 4(2)(1)$ $= 81 - 8$ $= 73$

UNDERSTANDING THE MAIN IDEAS

Using the quadratic formula

The quadratic formula is used to find the solutions of a quadratic equation.
Before using the quadratic formula, be sure that the quadratic equation is written
in the form $ax^2 + bx + c = 0$ so that the correct values of a, b, and c are used.

Example 1

Use the quadratic formula to find the solution(s) of the equation $2x^2 + 4x + 1 = 0$.

■ Solution ■

The equation is given in the required form: $a = 2$, $b = 4$, and $c = 1$. Using the
quadratic formula with these values,

$$x = \frac{-4 + \sqrt{4^2 - 4(2)(1)}}{2(2)} \qquad \text{and} \qquad x = \frac{-4 - \sqrt{4^2 - 4(2)(1)}}{2(2)}$$

$$= \frac{-4 + \sqrt{16 - 8}}{4} \qquad\qquad\qquad = \frac{-4 - \sqrt{16 - 8}}{4}$$

$$= \frac{-4 + \sqrt{8}}{4} \qquad\qquad\qquad\quad = \frac{-4 - \sqrt{8}}{4}$$

$$\approx -0.29 \qquad \leftarrow \text{Use a calculator.} \rightarrow \qquad \approx -1.71$$

The solutions of the equation are $x \approx -1.71$ and $x \approx -0.29$.

Find the solution(s) of each quadratic equation. Use the quadratic formula.

1. $3x^2 - 4x - 6 = 0$ **2.** $-x^2 + 3x - 2 = 0$ **3.** $x^2 + 12x + 10 = 0$

4. $2x^2 + 9x + 3 = 0$ **5.** $5x^2 - 20x = 0$ **6.** $3x^2 + 8x - 7 = 0$

7. $4x^2 - 20x + 25 = 0$ **8.** $3x^2 = 4x + 5$ **9.** $9x^2 + 12x + 4 = 0$

Value of the discriminant

The value of the discriminant, $b^2 - 4ac$, provides information about the number
of possible solutions of a quadratic equation.

If $b^2 - 4ac > 0$,   If $b^2 - 4ac = 0$,   If $b^2 - 4ac < 0$,
there are 2 solutions.   there is 1 solution.   there is no solution.

Note: Be sure the quadratic equation is in the form $ax^2 + bx + c = 0$ before
finding the value of the discriminant.

Example 2

Tell whether each equation has *two solutions, one solution,* or *no solution.*

a. $2x^2 + 6x + 5 = 0$ **b.** $x^2 - 6x + 9 = 0$ **c.** $2x^2 - 1 = 0$

■ Solution ■

a. The equation is given in the required form, with $a = 2$, $b = 6$, and $c = 5$.

$$b^2 - 4ac = (6)^2 - 4(2)(5)$$
$$= 36 - 40$$
$$= -4$$

Since $b^2 - 4ac < 0$, there is no solution of the equation.

b. The equation is given in the required form, with $a = 1$, $b = -6$, and $c = 9$.

$$b^2 - 4ac = (-6)^2 - 4(1)(9)$$
$$= 36 - 36$$
$$= 0$$

Since $b^2 - 4ac = 0$, there is one solution of the equation.

c. Write the equation as $2x^2 + 0x - 1 = 0$. So, $a = 2$, $b = 0$, and $c = -1$.

$$b^2 - 4ac = (0)^2 - 4(2)(-1)$$
$$= 0 - (-8)$$
$$= 8$$

Since $b^2 - 4ac > 0$, there are two solutions of the equation.

Tell whether each equation has *two solutions, one solution,* or *no solution.*

10. $x^2 - 3x - 18 = 0$ **11.** $x^2 - 2x + 1 = 0$ **12.** $3x^2 + 12 = 0$

13. $3x^2 - 10x + 3 = 0$ **14.** $2x^2 - 5x = 0$ **15.** $8x^2 - 10x + 5 = 0$

Finding the x-intercepts of a quadratic function

Since the x-intercepts of the graph of the quadratic function $y = ax^2 + bx + c$ occur where $y = 0$, the values of these x-intercepts are the solutions of the related quadratic equation $ax^2 + bx + c = 0$. As you know, there can be two, one, or no solutions of this related equation. When there are two solutions, the x-intercepts are $\dfrac{-b + \sqrt{b^2 - 4ac}}{2a}$ and $\dfrac{-b - \sqrt{b^2 - 4ac}}{2a}$. When there is one solution, the x-intercept of the graph is $\dfrac{-b}{2a}$.

Example 3

Find the x-intercepts, if any, for the graph of the function $y = x^2 - 3x - 18$.

■ Solution ■

If the graph of $y = x^2 - 3x - 18$ has x-intercepts, they occur where $y = 0$.

The value of the discriminant for the related equation $x^2 - 3x - 18 = 0$ is $b^2 - 4ac = (-3)^2 - 4(1)(-18) = 81$, so there are two x-intercepts.

The x-intercepts are $\dfrac{-b + \sqrt{b^2 - 4ac}}{2a}$ and $\dfrac{-b - \sqrt{b^2 - 4ac}}{2a}$, where $a = 1$, $b = -3$, and $c = -18$.

$$\frac{-(-3) + \sqrt{(-3)^2 - 4(1)(-18)}}{2(1)} \qquad \frac{-(-3) - \sqrt{(-3)^2 - 4(1)(-18)}}{2(1)}$$

$$= \frac{3 + \sqrt{9 + 72}}{2} \qquad\qquad = \frac{3 - \sqrt{9 + 72}}{2}$$

$$= \frac{3 + \sqrt{81}}{2} \qquad\qquad = \frac{3 - \sqrt{81}}{2}$$

$$= \frac{3 + 9}{2} \qquad\qquad\qquad = \frac{3 - 9}{2}$$

$$= 6 \qquad\qquad\qquad\qquad = -3$$

The x-intercepts are -3 and 6.

Find the x-intercepts, if any, for the graph of each function.

16. $5x^2 + 4x - 1 = 0$ **17.** $3x^2 - 15 = 0$ **18.** $8x^2 - 10x = -5$

· · · · · · · · · · · · · · · · · · · ·
Spiral Review

Use the method of completing the square to find the vertex of the graph of each equation. *(Section 5.4)*

19. $2x^2 + 4x + 1 = 0$ **20.** $x^2 - 6x + 9 = 0$ **21.** $2x^2 + 2x + 1 = 0$

Solve each equation. *(Section 5.1)*

22. $4x^2 - 64 = 0$ **23.** $\dfrac{1}{4}x^2 + 7 = 16$

Solve each equation. Round your answers to the nearest hundredth. *(Section 4.4)*

24. $3^{2x} = 15$ **25.** $\log_3 (x + 1) - \log_3 (x - 1) = 1$

Factoring Quadratics

Learn how to . . .

- factor quadratic expressions

- solve quadratic equations using factoring

So you can . . .

- solve equations by factoring

Application

On wet concrete, the distance, d, in feet needed to stop a car traveling at f miles per hour is given by the function $d = 0.055f^2 + 1.1f$. To find the speed that allows you to stop in 132 ft, solve $132 = 0.055f^2 + 1.1f$.

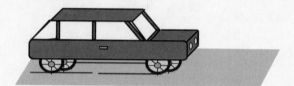

Terms to Know

Example / Illustration

Factoring (p. 222)	$x^2 + 4x + 3 = (x + 1)(x + 3)$
for an expression $x^2 + bx + c$, the process of finding (if possible) integers m and n such that $(x + m)(x + n) = x^2 + bx + c$	

UNDERSTANDING THE MAIN IDEAS

Factoring trinomials of the form $x^2 + bx + c$

To factor a trinomial of the form $x^2 + bx + c$ into the product of two binomials $(x + m)(x + n)$, try various combinations of the constants m and n in an attempt to find a product that results in the correct linear term of the trinomial. Since $(x + m)(x + n) = x^2 + (m + n)x + mn$, it follows that the correct combination is the one in which $m + n = b$ and $mn = c$.

Example 1

Write each quadratic function as a product of factors.

a. $y = x^2 + 5x + 6$ **b.** $y = x^2 - 7x + 10$ **c.** $y = x^2 + 4x - 12$

■ Solution ■

a. You need two integers m and n such that $m + n = 5$ and $mn = 6$. That is, you need two integers whose product is 6 and whose sum is 5. The required integers are 2 and 3.

$$y = x^2 + 5x + 6$$
$$= (x + 2)(x + 3) \qquad \leftarrow \text{Let } m = 2 \text{ and } n = 3.$$

The factorization of $y = x^2 + 5x + 6$ is $y = (x + 2)(x + 3)$.

b. Notice that the coefficient of the linear term is negative. You need two integers m and n such that $m + n = -7$ and $mn = 10$. That is, you need two integers whose product is 10 and whose sum is -7. The required integers are -5 and -2.

$$y = x^2 - 7x + 10$$
$$= (x + (-5))(x + (-2)) \qquad \leftarrow \text{Let } m = -5 \text{ and } n = -2.$$
$$= (x - 5)(x - 2)$$

The factorization of $y = x^2 - 7x + 10$ is $y = (x - 5)(x - 2)$.

c. Notice that the coefficient of the constant term is negative. You need two integers m and n such that $m + n = 4$ and $mn = -12$. That is, you need two integers whose product is -12 and whose sum is 4. (Recall that for the product of two integers to be negative, the integers must have different signs.) After trying various combinations, the required integers are -2 and 6.

$$y = x^2 + 4x - 12$$
$$= (x + (-2))(x + 6) \qquad \leftarrow \text{Let } m = -2 \text{ and } n = 6.$$
$$= (x - 2)(x + 6)$$

The factorization of $y = x^2 + 4x - 12$ is $y = (x - 2)(x + 6)$.

Write each quadratic function as a product of factors.

1. $y = x^2 + 9x + 14$ **2.** $y = x^2 + 3x + 2$ **3.** $y = x^2 - 10x - 24$

4. $y = x^2 - 6x + 8$ **5.** $y = x^2 + 13x - 30$ **6.** $y = x^2 + 2x - 15$

7. $y = x^2 + 5x - 14$ **8.** $y = x^2 - 6x + 5$ **9.** $y = x^2 - 14x + 24$

Factoring trinomials of the form $ax^2 + bx + c$

To factor $ax^2 + bx + c$ when $a \neq 1$, you need to find (if possible) integers $k, l, m,$ and n where:

$$ax^2 + bx + c = (kx + m)(lx + n)$$
$$= klx^2 + knx + lmx + mn$$
$$= klx^2 + (kn + lm)x + mn$$

While obviously more complicated than factoring a trinomial in which $a = 1$, you begin in a similar fashion. You again need to find two integers m and n whose product is c; however, you also need to find two integers k and l whose product is a. To be the required group of four integers, the expression $kn + lm$ must equal b, the coefficient of the linear term of the trinomial. A process called *guess and check* is often used.

Example 2

Find the x-intercepts of the graph of the function $y = 2x^2 + 9x + 10$.

■ Solution ■

Recall that the x-intercepts of a function can be found by solving the related equation formed when y is replaced by 0. So, the x-intercepts of $y = 2x^2 + 9x + 10$ can be found by solving the equation $2x^2 + 9x + 10 = 0$.

If the left side of the equation $2x^2 + 9x + 10 = 0$ can be factored, then the zero-product property can be used to find the solutions of the equation.

In the trinomial $2x^2 + 9x + 10$, $a = 2$ and $c = 10$. The only factors of 2 are 1 and 2, while the possible factors of 10 are 1 and 10, and 2 and 5. Check the value of $kn + lm$ for each combination of factors to find the required value, 9 (the value of b in the trinomial).

1 and 2, and 1 and 10: $1(1) + 2(10) = 21$ No

$\qquad\qquad\qquad\qquad 1(10) + 2(1) = 12$ No

1 and 2, and 2 and 5: $1(2) + 2(5) = 12$ No

$\qquad\qquad\qquad\qquad 1(5) + 2(2) = 9$ Yes   ($k = 1$, $n = 5$, $l = 2$, $m = 2$)

$\qquad 2x^2 + 9x + 10 = 0$

$\qquad (x + 2)(2x + 5) = 0$ ← Let $k = 1$, $n = 5$, $l = 2$, and $m = 2$.

By the zero-product property,

$\qquad x + 2 = 0 \qquad$ or $\qquad 2x + 5 = 0$

$\qquad\quad x = -2 \qquad\qquad\qquad\quad x = -2.5$

The x-intercepts of the graph of $y = 2x^2 + 9x + 10$ are $x = -2.5$ and $x = -2$.

Find the x-intercepts of the graph of each function.

10. $y = x^2 - 14x + 24$ **11.** $y = x^2 + 11x + 24$ **12.** $y = x^2 + 8x + 15$

13. $y = 3x^2 + 5x + 2$ **14.** $y = 5x^2 - 11x + 2$ **15.** $y = 3x^2 + 11x + 10$

16. $y = 4x^2 + 25x + 6$ **17.** $y = 10x^2 - 27x + 5$ **18.** $y = 4x^2 + 11x + 6$

Solve each equation. *(Section 5.5)*

19. $x^2 - 9x + 4 = 0$ **20.** $x^2 - 9x - 81 = 0$ **21.** $-3x^2 + 4x + 20 = 0$

Sketch the graph of each function. *(Section 5.2)*

22. $y = (x - 4)^2 + 5$ **23.** $y = -(x - 1)^2 + 3$ **24.** $y = 2(x - 3)^2 - 2$

Chapter 5 Review

Complete these exercises for a review of Chapter 5. If you have difficulty with a particular problem, review the indicated section.

For Exercises 1–3, solve each equation. Give solutions to the nearest tenth when necessary. *(Section 5.1)*

1. $x^2 + 2 = 11$
2. $-x^2 + 12 = -24$
3. $28 = 4x^2$

4. The graph of an equation of the form $y = ax^2$ passes through the point (2, 16). Find the value of a. *(Section 5.1)*

What is the maximum value or minimum value of each function? *(Section 5.2)*

5. $f(x) = -\dfrac{1}{2}(x + 2)^2 + 6$
6. $f(x) = 2x^2 - 8$
7. $f(x) = -2(x - 1)^2 + 10$

Write an equation in the form $y = a(x - h)^2 + k$ for each parabola shown. *(Section 5.2)*

8.

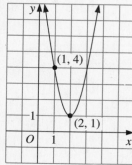

9.

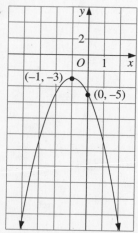

The equation $h = -16(t - 0.5)^2 + 24$ describes the path of a ball that is tossed straight up into the air, where h = the height (in feet) above the ground and t = the time (in seconds) since it was tossed. *(Section 5.2)*

10. After how many seconds will the ball reach a maximum height?

11. What is the maximum height that the ball will reach?

For each equation:

a. Find the x-intercept(s).

b. Find the vertex.

c. Sketch the graph. *(Section 5.3)*

12. $y = 3(x - 4)(x + 1)$
13. $y = -2(x - 1)(x + 3)$

Write an equation in intercept form for each graph. *(Section 5.3)*

14.

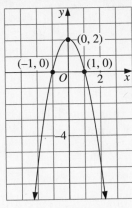

15.

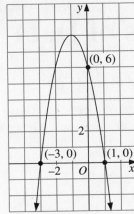

Write each function in vertex form. *(Section 5.4)*

16. $y = -x^2 + 6x + 3$ **17.** $y = 2x^2 - 6x + 2$ **18.** $y = -3x^2 - 12x + 2$

Find the solution(s) of each quadratic equation. Use the quadratic formula.
(Section 5.5)

19. $x^2 - 10x + 8 = 0$ **20.** $3x^2 - 27x = 0$ **21.** $2x^2 + 8x = -6$

Find the x-intercepts, if any, for the graph of each quadratic function.
(Section 5.5)

22. $y = -x^2 + 6x + 3$ **23.** $y = 2x^2 - 10x + 13$

Write each quadratic function as a product of factors. *(Section 5.6)*

24. $y = x^2 - 7x + 6$ **25.** $y = x^2 - x - 20$ **26.** $y = 3x^2 - 11x + 10$

SPIRAL REVIEW   Chapters 1–5

Rewrite each expression in factored form.

1. $35x - 14$ **2.** $7x^2 + 14x$

Solve each equation. Round your answers to the nearest hundredth.

3. $4^x = 18$ **4.** $\log_3 (x + 1) + \log_3 (x - 1) = 2$

Sketch the graph of each function.

5. $y = (x - 4)^2 + 5$ **6.** $y = -(x - 1)^2 + 3$ **7.** $y = 2(x - 3)^2 - 2$

Find the inverse of each function.

8. $f(x) = 3x + 1$ **9.** $f(x) = 3.2x - 1.4$ **10.** $f(x) = 4^x$

Write as a logarithm of a single number or expression.

11. $-\dfrac{1}{2} \log_b 16$ **12.** $3 \log_b m + 4 \log_b n$ **13.** $2 \log_b y - 3 \log_b y$

6.1

Types of Data

Learn how to . . .

- organize information and classify data

So you can . . .

- analyze data

Application

Baseball teams collect data about wins and losses, hits, runs scored, home runs, pitchers' records, and many other types of data, in order to plan their strategies for each game.

Terms to Know

Example / Illustration

Terms to Know	Example / Illustration
Numerical data (p. 238) counts or measurements	The attendance figures at a baseball stadium for each home game are numerical data.
Categorical data (p. 238) names or labels	The days of the week that the home team won during this week are categorical data.
Population (p. 239) a complete group	All of the students in your high school are a population.
Sample (p. 239) part of a group	Just the students in your algebra class are a sample of all the students in your high school.

UNDERSTANDING THE MAIN IDEAS

When analyzing data, consider these two distinct data types: numerical and categorical.

Example 1

Tell whether the data that can be gathered about each variable are *categorical* or *numerical*. Then describe the categories or numbers.

  a. favorite pizza toppings

  b. prices of medium pizzas in your town

■ Solution ■

  a. The data are categorical. The toppings might be mushrooms, onions, olives, green peppers, and so on.

  b. The data are numerical. The prices are monetary amounts in dollars and cents.

Tell whether the data that can be gathered about each variable are *categorical* or *numerical*. Then describe the categories or numbers.

1. a person's favorite juice flavor

2. the home runs hit by a baseball team each month of the season

3. the points scored on the floor exercise at a gymnastics meet

4. the types of books read by your classmates

To analyze the data for any situation, it is important to consider the population and the type of display that will provide the best view of the information. The display can then be used to efficiently answer any question about the data.

Example 2

Your class wants to compare the prices of a medium pizza that could be delivered.

  a. Describe the population and the sample.

  b. Tell what type of data would be collected.

  c. Tell what type of graph you would use to display the data.

■ Solution ■

  a. The population consists of all the local restaurants that deliver pizzas. The sample is those restaurants that the students call to ask for prices.

  b. The collected data will be numerical.

  c. The data might be displayed in a histogram, with price intervals along the horizontal axis and frequency along the vertical axis.

For Exercises 5–9:

a. Describe the population and the sample.

b. Tell what type of data would be collected.

c. Tell what type of graph you would use to display the data.

5. A tire manufacturer tests 100 tires of a certain type to determine the life of the tire in miles.

6. The dance committee at your school wants to know how much to charge for tickets. They survey students from each grade level, asking the maximum price the students would pay for a ticket.

7. The recycling company in a certain city wants to know how many bins to place at each collection site. They survey the people in each neighborhood, asking about the amount of material they recycle each week.

8. A candidate for President of the Student Council of your school wants to know what issues are most important to her fellow students. She telephones students from each grade level to ask about their ideas for changes in various school practices.

9. The librarian at your school wants to plan the orders for next year's magazine subscriptions. He surveys the students who enter the library, asking them what magazines they read regularly.

· · · · · · · · · · · · · · · · · · · ·

Spiral Review

Find the *x*-intercept(s) of the graph of each function. *(Section 5.6)*

10. $y = 2x^2 - x - 1$

11. $y = 6x^2 + 5x + 1$

12. $y = x^2 - 7x + 12$

13. $y = x^2 - 5x - 24$

Graph each equation. *(Section 5.3)*

14. $y = 2(x - 1)(x + 2)$

15. $y = (x - 4)(x + 4)$

16. $y = -(x - 6)(x + 3)$

17. $y = 3(x - 2)(x + 5)$

6.2

Writing Survey Questions

Learn how to . . .

- write good survey
  questions

So you can . . .

- get more accurate
  responses to your
  surveys

- judge the accuracy
  of responses to
  other people's
  surveys

Application

Candidates for political office often have committees that survey the voters in
order to decide which issues are most important, and what position they should
take on individual questions.

Terms to Know

Example / Illustration

Biased question (p. 245)	"Global warming is increasingly
a question that produces responses that do not accurately reflect the opinions or actions of the respondents	responsible for unusual weather patterns. Do you favor government spending to reduce global warning?"

UNDERSTANDING THE MAIN IDEAS

Biased questions

To get useful and accurate responses to a survey, it is important to ask questions
that are not biased. Biased questions may lead the respondent to a certain
response or may not provide them with enough information to give an accurate
opinion.

Example 1

Tell why the following question may be biased. Then describe what changes you
would make to improve the question.

"Do you favor a proposal to increase spending on technology for our schools?"

■ Solution ■

This question assumes that the respondent is familiar with the proposal. So the responses to the question by people not totally familiar with the proposal could result in conclusions that are inaccurate.

The question could be improved by mentioning the specifics of the proposal, including the amount of the increase.

For Exercises 1–6, tell why each question may be biased. Then describe what changes you would make to improve the question.

1. "What grade did you receive in your last math class?"

2. "A survey shows that only 10% of the voters favor a tax increase. Do you favor a tax increase?"

3. "Do you think the election of class officers is done fairly?"

4. "Are you in favor of destroying old houses that need repairs?"

5. "Which of the following soups should be eliminated from the menu in the school cafeteria?"

   A. tomato   **B.** chicken and rice   **C.** vegetable

6. "How much weight have you gained during the past year?"

Format of a survey

Some surveys have respondents choose their answers to a question from a list of options. Surveyors have found that results are more accurate when people are given a list of choices as opposed to an open-ended question. A list of choices makes it more likely that people will complete the survey, and also makes it easier for the surveyor to analyze the results.

Example 2

A survey sent out by a local radio station asked the following questions.

1. How often do you listen to station WXYZ?

2. Do you listen to the station because of the music or the local news?

3. Would you listen to the station if there were no news at all?

4. What kind of music would you like to hear more often?

Rewrite the survey questions to include options for each question that the respondents could choose.

■ Solution ■

Place a check mark next to your response.

1. How often do you listen to station WXYZ?
less than 1 hour a week ___   1–3 hours a week ___
4–6 hours a week ___   more than 7 hours a week ___

2. Do you listen to the station because of the music or the local news?
music ___   local news ___

3. Would you listen to the station if there were no news at all?
yes ___   no ___

4. What kind of music would you like to hear more often?
rock ___   country ___   jazz ___   oldies ___

Use the survey shown above to answer Exercises 7–9.

7. What information is the station trying to obtain?

8. Which questions are worded well?

9. Explain how the data from this survey could be used to increase the audience for station WXYZ.

10. Write a survey that includes at least three questions that would help the director of transportation in a city create a bus schedule that best serves the public's needs.

· · · · · · · · · · · · · · · · · · ·
Spiral Review

Tell whether the data that can be gathered about each variable are *categorical* or *numerical*. Then describe the categories or numbers. *(Section 6.1)*

11. the type of transportation a student uses to get to school

12. the amount of time a student spends on homework each day

13. a person's height in inches

Match each scatter plot to one of the following values of the correlation coefficient *r*: –0.55, 0.88, 0.3. *(Section 2.5)*

14. 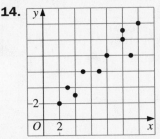 **15.** 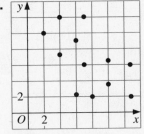 **16.**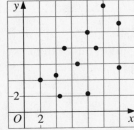

Collecting Data from Samples

Learn how to . . .

- choose a representative sample

So you can . . .

- recognize biased samples in surveys

Application

Before opening a new store, the owners of Happy Harry's surveyed 50 homes in the neighborhood around the new location to find out whether people are satisfied with the stores they now shop, and what improvements they would like to see.

Terms to Know	*Example / Illustration*
Self-selected sample (p. 250) data collected by letting people volunteer	A survey form is available in a small store, asking shoppers to mail in their suggestions or improvements.
Systematic sample (p. 250) data collected by choosing people using a pattern	Survey every fourth person who walks through the door.
Convenience sample (p. 250) data collected by choosing people who are easy to reach	Survey the people in line at the cash register in the front of the store.
Random sample (p. 250) choosing people so that each person has an equally likely chance of being selected	Survey people selected at random from the phone book.
Stratified random sample (p. 250) choosing people by dividing the population into groups and then randomly selecting people from each group	Survey random selections of women and random selections of men.
Cluster sample (p. 250) choose people as a group rather than as individuals	Survey a group of people in the parking lot of a shopping center.

Terms to Know	Example / Illustration
Biased sample (p. 251) a sample that overrepresents or underrepresents part of the population	A sample of only females at a high school on cafeteria food will not represent the opinions of male students.

UNDERSTANDING THE MAIN IDEAS

Sampling a population

To obtain useful information from a survey when the entire population cannot be surveyed, it is important to consider the method for choosing a sample of the population so that it will represent the population accurately. There are six sampling methods: self-selected, systematic, convenience, random, stratified random, and cluster. A random sample is preferred since it is most likely to produce a representative sample of the population.

Example 1

A car dealer wants to know what color people prefer when ordering a new car. He asks every visitor to the showroom on Sunday afternoon what color they would choose for their next new car. Is this a *random sample*, a *cluster sample*, a *convenience sample*, or a *stratified random sample*? How representative of all new car buyers do you think the sample will be?

▪ Solution ▪

The sample is a convenience sample because only those people how came into the showroom were surveyed. Also, since the survey was done on just one day, it is not likely to be a representative sample of all new car buyers.

In Exercises 1–5, four students were planning to survey ten consumers about their preferences of frozen yogurt flavors. For each student's plan, identify the type of sample and describe the population of the survey.

1. Jennifer will ask one child and one adult from each of the first five families who enter a store called the YogurtShop about their preferences.

2. Seema will ask the first ten students who come into her algebra class.

3. Shelton will survey ten people chosen at random from the phone book.

4. Anil will ask each of the students seated at two of ten five-person tables in the school lunchroom.

5. Explain how you would choose the ten people.

Biased samples

A biased sample does not accurately represent the members of a population. Such a sample places too much or too little emphasis on a part or parts of a population.

Example 2

Identify the type of sample and describe the population of the survey. Then tell if the sample is biased. Explain your reasoning.

There are 300 students at Plainville High School, of whom 30% are freshmen, 25% are sophomores, 25% are juniors, and 20% are seniors. To estimate the number of hours of television the students watch, a Social Studies teacher numbers the students in each grade level beginning with 1 and then generates a random selection that includes 25 freshmen, 25 sophomores, 25 juniors, and 25 seniors.

■ Solution ■

The sample is a stratified random sample and the population is the entire student body of Plainville High School. However, the sample is biased because there are an equal number of students surveyed from each grade level, while the number of students in each grade level is not the same. It would be better to randomly select 30 freshmen, 25 sophomores, 25 juniors, and 20 seniors.

Identify the type of sample and describe the population of the survey. Then tell if the sample is biased. Explain your reasoning.

6. A grocery store manager wants to survey customers regarding their preferences about the size of milk containers. He surveys 20 customers at the front door of the store.

7. A citizens' committee in a small town wants to survey the residents regarding their attitudes about their state government raising the minimum driving age to 18. They mail surveys to all the homeowners in the community.

8. A local school board wants to find out how the school district's patrons feel about an increase in the school tax. They select names at random from the telephone book and each member calls 25 of these people.

9. A political candidate wants to know how the voters feel about recycling activities. She surveys people on Saturday morning at the local recycling center.

For Exercises 10–12:

a. **Describe any bias in the question.**

b. **Tell whether the data that can be gathered about each variable are *categorical* or *numerical*. Then tell what type of graph you might use to display the results.** *(Section 6.1)*

10. "Do you think it is important for seniors to spend all day in school?"

11. "What do you think is the minimum number of hours that a student should spend on homework?"

12. "Do you support the school board's decision to add an extra hour to the school day?"

13. Find the mean, the median, and the mode of the following temperature data that was collected in Wilmington in one week in March.

Day of week	High temperature (°F)
Monday	42
Tuesday	38
Wednesday	56
Thursday	56
Friday	60
Saturday	45
Sunday	48

Displaying and Analyzing Data

Learn how to . . .

- make histograms and box plots

So you can . . .

- create appropriate displays for numerical data

Application

The heights of the ten basketball players on the Sidney Celtics are 69 in., 74 in., 74 in., 76 in., 76 in., 76 in., 78 in., 79 in., 81 in., and 84 in. This data can be displayed using a histogram or a box plot.

Terms to Know	Example / Illustration
Box plot (p. 259) a data display that shows the median, the quartiles, and the extremes of a data set (The lower quartile is the median of the lower half of the data, the upper quartile is the median of the upper half of the data, and the lower and upper extremes are the least and greatest data values, respectively.)	**Heights of Sidney Celtics Players (in.)** 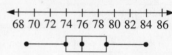 68 70 72 74 76 78 80 82 84 86 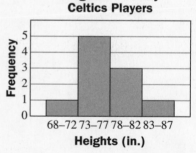
Histogram (p. 260) a graph that displays numerical data (The horizontal axis of a histogram is a number line divided into intervals of equal length. The vertical axis shows the frequency of the data items that fall within each interval.)	**Heights of Sidney Celtics Players**
Absolute frequency (p. 260) an actual count of the data within each interval	The absolute frequency for the heights in the interval 73–77 (as shown in the histogram above) is 5.
Relative frequency (p. 260) the percent of all the data that falls within an interval	The relative frequency for the heights in the interval 73–77 is $\frac{5}{10} = 0.5$, or 50%.

Terms to Know	Example / Illustration
Symmetric distribution (p. 260) a histogram with a shape that is approximately symmetrical about a vertical line passing through the interval with the greatest frequency	The histogram shown on the previous page is not a symmetric distribution because there are more players with heights less than 75 in. than there are players with heights greater than 79 in.
Skewed distribution (p. 260) a histogram with a "tail" in one direction	The distribution of heights of the Celtics' team members is skewed, with a "tail" to the right.

UNDERSTANDING THE MAIN IDEAS

A data display should present information in a form that is easily understood. It is important to know how to make a good display so that you present information accurately.

Example 1

Refer to the heights of the Sydney Celtics' basketball players given in the Application, as well as the box plot and histogram shown in the Terms to Know table.

   a. What is the median of the heights of the team members?

   b. Between what two numbers do 50% of the heights fall?

   c. Is the histogram for the heights of the players *symmetric* or *skewed*?

Solution

   a. The median height is 76 in.

   b. From the box plot, 50% of the heights fall between 74 in. and 79 in.

   c. The histogram is skewed, with a "tail" to the right.

Example 2

An airline conducted an audit of its fifteen flights leaving a busy national airport one day to learn how many minutes late each flight had taken off. The results (reported to the nearest 5 min) are shown below.

15, 10, 10, 40, 15, 0, 20, 10, 0, 0, 25, 5, 20, 5, 10

Display the results in a box plot.

Solution

Step 1: Find the median, the quartilies, and the extremes of the data.

Write the data values in order from least to greatest, and then identify the key values.

0, 0, 0, 5, 5, 10, 10, 10, 10, 15, 15, 20, 20, 25, 40
↑ ↑ ↑ ↑ ↑
lower   lower   median   upper   upper
extreme quartile   quartile extreme

Step 2: Make the box plot by drawing a box from the lower quartile to the upper quartile, line segments from the box to the extremes, and a vertical line through the median.

Departure Delays on an Airline's Flights (min)

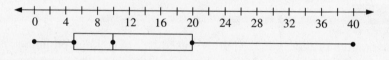

For Exercises 1–10, use the following data, which are the heights of the members of the Townsend Tigers' basketball team.

69 in., 69 in., 70 in., 72 in., 74 in., 74 in., 75 in., 75 in., 79 in., 84 in.

1. Find the median height for the members of this team.

2. Find the lower quartile and the upper quartile of the data.

3. Make a box plot of this data.

4. Between what two numbers do 50% of the data fall?

5. Below what number do 50% of the data fall?

6. What are the lower extreme and upper extreme of the data?

7. How do the extremes of the heights of the Townsend Tigers compare with the extremes of the heights of the Sidney Celtics?

8. Compare the box plot of the heights of the Tigers' team members with the box plot of the heights of the Celtics' team members. Which team has a height advantage? Explain your answer.

9. Make a histogram for the heights of the Tigers' team members. Use the same intervals as the heights of the Celtics' team members. Is the histogram of heights of the Tigers' team members symmetric? Explain.

10. Which display tells you more about the individual players on each team, a histogram or a box plot? Explain your answer.

········

Spiral Review

Identify the type of sample described. Tell if the sample is biased. Explain your reasoning. *(Section 6.3)*

11. The local school board is deciding whether to raise the prices of tickets to basketball games. They instruct the school administration to survey every fourth person who attends the next Saturday night home basketball game regarding the proposed new ticket prices.

12. A manager of a radio station wants to know if the station's listeners want to hear more country music. The manager has the disc jockeys ask their listeners several times in one day to call in with their votes.

Simplify. *(Toolbox, page 777)*

13. $\sqrt{32}$ 14. $\sqrt{300}$ 15. $\sqrt{\dfrac{75}{3}}$ 16. $\sqrt{\dfrac{336}{3}}$

Describing the Variation of Data

Learn how to . . .

- find and interpret the range, interquartile range, and standard deviation of a data set

So you can . . .

- determine the variability of data

Application

A patient's blood pressure can vary from day to day, as well as from the morning to the evening. A doctor will look at the individual reading, as well as the range of the readings and the variation of the readings to determine the health of a patient.

Terms to Know	*Example / Illustration*
Range (p. 266) the difference between the maximum and minimum data values	For data values 1, 1, 2, 4, 5, 6, and 6, the range is $6 - 1 = 5$.
Interquartile range (p. 266) the difference between the upper quartile and the lower quartile of a data set	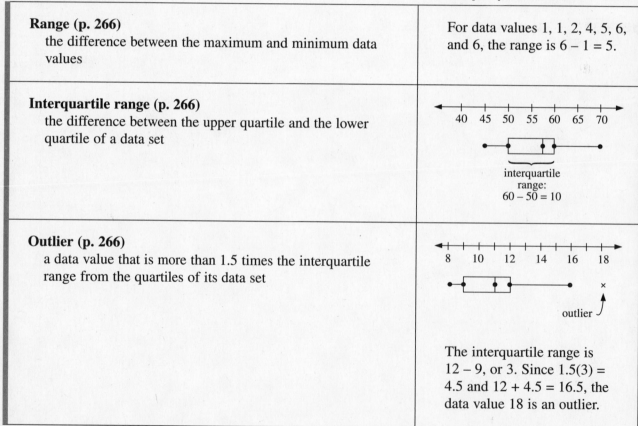
Outlier (p. 266) a data value that is more than 1.5 times the interquartile range from the quartiles of its data set	The interquartile range is $12 - 9$, or 3. Since $1.5(3) = 4.5$ and $12 + 4.5 = 16.5$, the data value 18 is an outlier.

Terms to Know	Example / Illustration
Standard deviation (p. 268) a measure of variability that uses the square of the deviation of every data value from the mean (The symbol for standard deviation is the lowercase Greek letter σ.)	The standard deviation of the data values 1, 1, 2, 4, 5, 6, and 6 is about 2.06.

UNDERSTANDING THE MAIN IDEAS

When a data set is used as the basis for conclusions, it is important to know the individual data values, the mean of the data, and how much the data vary. To determine the variability of a set of data, it is often helpful to find the range, the interquartile range, and any outliers of the data set.

Example 1

The owner of a city parking lot gathers data on the number of cars in the lot at 9 A.M. each day for two weeks. The following box plot shows a display of the collected data.

Number of Cars in Lot at 9 A.M.

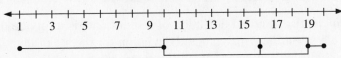

   1   3   5   7   9   11   13   15   17   19

  a. Find the range of the data.

  b. Find the interquartile range of the data

  c. Find any outliers.

■ Solution ■

  a. Since the greatest data value is 20 and the least value is 1, the range of the data set is 20 – 1, or 19.

  b. From the box plot, the upper quartile value is 19 and the lower quartile value is 10, so the interquartile range is 19 – 10, or 9.

  c. Using the interquartile range, 9, from part (b), if there are any outliers they must be 1.5(9), or 13.5 units less than the lower quartile value or 13.5 units greater than the upper quartile.

$$10 - 13.5 = -3.5 \qquad\qquad 19 + 13.5 = 32.5$$

Since the lower extreme of the data set is 1 and the upper extreme is 20, there are no outliers.

The owner of a city parking lot gathers data on the number of cars in the lot at 11 A.M. each day for two weeks. The following box plot shows a display of the data collected. Use the box plot to find each value for the data.

Number of Cars in Lot at 11 A.M.

1. Find the range of the data.

2. Find the interquartile range of the data.

3. Identify any outliers in the data.

Standard deviation

To find the standard deviation of a set of data, you first find the mean of the data. Then for each data value, you find the difference between the data value and the mean. Next you square each of the differences found and then find the sum of these squares. Finally, you divide this sum by the number of data values and take the square root of this quotient. A calculator with statistical functions can be used to compute the standard deviation of a set of data.

Example 2

One way to *estimate* the standard deviation of a set of data is to divide the range by 4. Use this rule to estimate the standard deviation for the set of data below. Then find the actual standard deviation and compare your results.

The systolic blood pressure readings for a patient for 14 consecutive days were 170, 176, 180, 180, 176, 163, 147, 168, 138, 173, 154, 158, 170, and 166.

■ Solution ■

The highest reading is 180 and the lowest is 138. So the range is $180 - 138$, or 42. The rule gives an estimated standard deviation of $42 \div 4$, or 10.5.

Using a calculator or statistical software, the standard deviation, σ, is about 12.1.

The calculated value for the standard deviation is somewhat higher than the estimated value.

One way to estimate the standard deviation of a set of data is to divide the range by 4. For Exercises 4–7, use this rule to estimate the standard deviation for each set of data. Then find the standard deviation and compare your results.

4. 9, 6, 7, 4, 9

5. 8, 6, 6, 7, 4, 7, 4

6. 8, 14, 13, 15, 8

7. 3, 6, 12, 15

ree sets of data are shown below. Without calculating, which set will have
 greatest standard deviation? Which set will have the least standard
·iation? Explain your answers.

t A: 1, 2, 3, 4, 5   Set B: 1, 3, 3, 3, 5   Set C: 1, 1, 3, 5, 5

....
Spiral Review

·ealtor took a survey of the prices of houses offered for sale in her city.
e prices, in thousand of dollars, are given below. Make a histogram of the
a. Describe the distribution of the data. *(Section 6.4)*

5, 110, 105, 172, 209.5, 114, 148, 174.9, 139.9, 330, 159.9, 67.9, 100,
.9, 169.9, 139.9, 189.9, 375, 170.9, 179.9, 94.9, 319, 289.8, 185, 275

·ach inequality. *(Toolbox, p. 787)*

· 3 > 15   **11.** $-2x - 5 \leq 13$ **12.** $2x + 3 \geq 19$

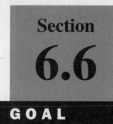

Learn how to . . .

- find the margin of error for a sample proportion

So you can . . .

- make decisions about opinion poll results

Making Decisions from Samples

Application

In a survey of 300 voters one day before an election for state senator, Coretta Kay received 153 votes. Her opponent says she can still be beaten. Do you think he can convince voters that he is right?

Terms to Know

Example / Illustration

Sample proportion (p. 273) the percent of responses to a survey that are given in a certain way	In the Application above, the sample proportion of votes for Coretta Kay is $\frac{153}{300} = 0.51$, or 51%.
Sampling distribution (p. 274) a display of many samples taken from the same population (The display is usually a histogram.)	The distribution of the sample was 51% for Coretta Kay and 49% for her opponent.
Sample size (p. 274) the number of responses to a question taken as a sample (The variable n is used to represent the sample size.)	The sample size for this survey was 300.
Margin of error (p. 274) the expected variability in a sampling (The margin of error for a sample proportion is approximated by the expression $\pm\frac{1}{\sqrt{n}}$.)	For the sampling discussed in the Application, the margin of error is $\pm\frac{1}{\sqrt{n}} = \frac{1}{\sqrt{300}}$ $\approx \pm 0.058$ $= \pm 5.8\%$

UNDERSTANDING THE MAIN IDEAS

By looking at the formula for computing the margin of error for the results of a survey, you can see that as the sample size n increases, the margin of error decreases. This simply means that as a surveyor collects more and more samples, the results more accurately reflect the opinions of the entire population.

Example 1

During a survey of 1000 Americans, the correct response to a question about a certain consumer product was given 35% of the time. Find the interval that is likely to contain the proportion of all Americans who would give the correct response to the question.

■ Solution ■

Since $n = 1000$, the margin of error is

$$\pm\frac{1}{\sqrt{n}} = \pm\frac{1}{\sqrt{1000}}$$

$$\approx 0.032$$

$$\approx 3.2\%$$

Therefore, the interval that is likely to contain the proportion of all Americans who would give the correct response to the question is $35\% \pm 3.2\%$, or from 31.8% to 38.2%.

During a survey of n Americans, the given sample proportion gave the correct response to a question about a certain consumer product. For each value of n and the given sample proportion, find the interval that is likely to contain the proportion of all Americans who would answer the question correctly.

1. $n = 500$; sample proportion: 67%

2. $n = 1200$; sample proportion: 85%

3. $n = 2000$; sample proportion: 25%

If you know the margin of error for a sample, you can find the size of the sample that was used.

Example 2

If the margin of error for an opinion poll is $\pm 1\%$, find the size of the sample. Why do polls rarely have a margin of error this small?

■ Solution ■

Use the margin of error formula and solve for n.

$$\pm 0.01 = \frac{1}{\sqrt{n}} \qquad \leftarrow \pm 1\% = \pm 0.01$$

$$0.01 = \frac{1}{\sqrt{n}}$$

$$0.01\left(\sqrt{n}\right) = \left(\frac{1}{\sqrt{n}}\right)\sqrt{n}$$

$$0.01\left(\sqrt{n}\right) = 1$$

$$\sqrt{n} = \frac{1}{0.01}$$

$$\sqrt{n} = 100$$

$$\left(\sqrt{n}\right)^2 = 100^2$$

$$n = 10{,}000$$

The sample size was 10,000. Polls rarely have a margin of error this small because of the large sample size that must be surveyed.

For Exercises 4–6, find the sample size that corresponds to the given margin of error.

4. $\pm 1.5\%$ **5.** $\pm 2\%$ **6.** $\pm 3\%$

7. The registered voters in a city were surveyed on their preference for a candidate for mayor. Of 1200 women surveyed, 50% said they would vote for the incumbent. When 1200 men were surveyed with the same question, 52% said they would vote for the incumbent. Is it reasonable to conclude that more men than women will vote for the incumbent? Explain.

· · · · · · · · · · · · · · · · · · ·
Spiral Review

Write a point-slope equation of the line passing through each pair of given points. *(Section 2.3)*

8. $(4, -1)$ and $(-1, -1)$ **9.** $(1, -3)$ and $(5, 10)$

10. $(-2, 5)$ and $(3, -1)$ **11.** $(-5, 1)$ and $(5, 11)$

Write an equation of each function in vertex form. *(Section 5.4)*

12. $y = -x^2 + 2x$ **13.** $y = x^2 - 4x + 3$

14. $y = x^2 - 2x + 2$ **15.** $y = -2x^2 - 8x - 3$

Chapter 6 Review

Complete these exercises for a review of Chapter 6. If you have difficulty with a particular problem, review the indicated section.

For Exercises 1 and 2, tell whether the data that can be gathered about each variable are *categorical* or *numerical.* Then describe the categories or numbers. *(Section 6.1)*

1. months of the year

2. grade point averages

3. Your algebra class is collecting data on their blood types in order to predict the proportions of each blood type for the entire school. Describe the population and the sample. *(Section 6.1)*

4. Tell what type of graph you would use to display the blood type data discussed in Exercise 3. *(Section 6.1)*

5. A survey asked the question "Do you believe that the city council should approve the proposed purchase of several downtown riverfront properties?" Is this a biased question? Explain your answer. *(Section 6.2)*

6. One question on a survey designed to discover a person's attitude about their job is: "Do you think you are paid fairly for all the hours you spend on your job?" Is the question biased? Explain your answer. *(Section 6.2)*

7. The editors of a local newspaper want to know how many people in their area read the sports section of the newspaper. They have their sports reporter survey people at a Saturday night basketball game. Identify the type of sample and describe the population. Then tell if the sample is biased. Explain your reasoning. *(Section 6.3)*

8. Of the 300 students at Plainville High School, 35% are freshmen, 25% are sophomores, 20% are juniors, and 20% are seniors. To estimate the number of hours of television watched by the average student, a sample of 100 students will be surveyed. Describe a method for selecting a stratified random sample. *(Section 6.3)*

The daily high temperature readings (in °F) in Kent during the first two weeks of May are given below. *(Section 6.4)*

70, 76, 80, 80, 76, 63, 47, 68, 38, 73, 54, 58, 70, 66

9. Make a histogram for the temperatures. Use the intervals 30–39, 40–49, 50–59, 60–69, 70–79, and 80–89.

10. Tell whether the distribution of temperatures is *symmetric* or *skewed.*

For Exercises 11–13, use the box plot below. *(Sections 6.4 and 6.5)*

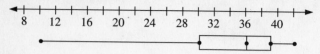

11. What are the range, the interquartile range, and the median of the data?

12. Identify any outliers.

13. Between what two numbers do 50% of the data fall?

For Exercises 14 and 15, use the temperature data from Questions 9 and 10. *(Section 6.5)*

14. Estimate the standard deviation of the data by dividing the range by 4.

15. Find the actual standard deviation of the data set.

16. During a survey of 1400 Americans, the correct response to a question about a certain consumer product was given 53% of the time. Find the interval that is likely to contain the proportion of all Americans who would answer correctly. *(Section 6.6)*

17. If the margin of error for a poll is ±4%, find the size of the sample. *(Section 6.6)*

SPIRAL REVIEW   Chapters 1–6

Graph each equation.

1. $y = (x - 1)(x + 3)$

2. $y = (x - 2)(x + 2)$

3. $y = -(x + 6)(x - 3)$

4. $y = 3(x - 1)(x + 4)$

For Exercises 5 and 6, find the x-intercepts of the graph of each function.

5. $y = 2x^2 - x - 1$

6. $y = 3x^2 - x - 4$

7. The high temperature readings (in °F) in Kent for fourteen consecutive days in May are given below. Find the mean, the median, and the mode of the data.

70, 76, 80, 80, 76, 63, 47, 68, 38, 73, 54, 58, 70, 66

Simplify.

8. $\sqrt{162}$

9. $\sqrt{500}$

10. $\sqrt{\dfrac{27}{3}}$

11. $\sqrt{\dfrac{36}{4}}$

Solve each inequality.

12. $x + 5 > 13$

13. $-2x + 5 \le 13$

14. $2x - 5 \ge -19$

Write a point-slope equation of the line passing through each pair of points.

15. $(4, -3)$ and $(-1, -3)$

16. $(-1, -3)$ and $(2, 6)$

Write each function in vertex form.

17. $y = x^2 - 4x + 2$

18. $y = -2x^2 - 6x - 3$

Systems of Equations

Application

A baseball player's batting average is given by the ratio

$$\frac{\text{number of hits}}{\text{number of official at-bats}}.$$

If you know the number of hits and at-bats a player has during a game, and if you know the player's batting average before and after the game, you can use a system of equations to find the player's total number of hits and total number of at-bats for the season.

Terms to Know

System of equations (p. 292)
   two or more equations involving the same variables

Example / Illustration

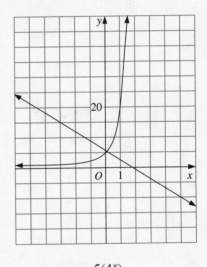

$y = 5(4^x)$
$y = 5.5 - 3.2x$

UNDERSTANDING THE MAIN IDEAS

Solving a system of equations means finding the coordinates of the point(s) where the graphs of the equations intersect. The coordinates of a point of intersection will satisfy all the equations of the system.

Example

To convert temperatures from the Celsius scale to the Fahrenheit scale, you can use the exact formula $F = \frac{9}{5}C + 32$ or you can use the approximation formula $F = 2C + 30$. Find the values of F and C for which the approximation is exact,

a. using a graphing calculator.

b. using substitution.

▪ Solution ▪

a. Enter the equations $y_1 = \frac{9}{5}x + 32$ and
$y_2 = 2x + 30$ into a graphing calculator.
Using the Trace feature, you can see that
the two graphs intersect at the point with
coordinates (10, 50).

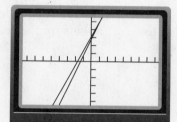

b. Since $F = \frac{9}{5}C + 32$, substitute $\frac{9}{5}C + 32$ for
F in the formula $F = 2C + 30$.

$$F = 2C + 30$$

$$\frac{9}{5}C + 32 = 2C + 30$$

$$32 - 30 = 2C - \frac{9}{5}C$$

$$2 = \frac{1}{5}C$$

$$C = 10$$

Now use this value of C to find the corresponding value of F.

$$F = \frac{9}{5}C + 32$$

$$F = \frac{9}{5}(10) + 32$$

$$= 18 + 32$$

$$= 50$$

The answer is $C = 10$ and $F = 50$.

Find the point of intersection of each pair of lines.

1. a vertical line through (5, 0) and a horizontal line through (0, 1)

2. a vertical line through (−3, 1) and a horizontal line through (−1, −1)

3. the x-axis and a vertical line through (6, 8)

Use a graphing calculator or graphing software to solve each system of equations. (The graphs of the equations may intersect more than once.)

4. $y = 3x^2$
$y = 3(2^x)$

5. $xy = 1$
$y = x$

6. $y = x + 5$
$y = x - 1$

7. $y = 7$
$x + y = 2$

For Exercises 8–11, use substitution to solve each system of equations.

8. $y = 3x$
$2x + y = 10$

9. $y = x - 1$
$5x - y = 13$

10. $3x + y = 1$
$x - 2y = -9$

11. $4x + 6y = 3$
$x - 8y = -4$

12. At the end of last week, a baseball player's batting average was .256. One week later, after getting 12 hits in 35 at-bats, the player's average is now .275. How many hits and how many at-bats does the player now have for the season?

························
Spiral Review

For Exercises 13–15, solve each system of equations. If all numbers are solutions, write *all numbers*. If there is no solution, write *no solution*.
(Toolbox, page 786)

13. $3x - 5 = 13 - 3x$

14. $7x + 11 = -4 + 7x$

15. $4x + 9 = 2x - 7$

16. a. Make a scatter plot of the ordered pairs (2, 0), (5, 3), (–1, –3), and (0, –2). *(Section 4.5)*

b. If the four points represent the ordered pair (log m, log t), what is an equation relating m and t?

c. What is the value of t when $m = 500$?

GOAL

Learn how to . . .

- solve linear systems by adding equations

So you can . . .

- solve problems involving two unknown values

Linear Systems

Application

On a computer disk for a school newspaper, each graphic requires one amount of storage space and each text file requires another amount of storage space. You can use a linear system of equations to describe the combinations of graphics and text files that can be stored on the disk.

Terms to Know

Inconsistent system (p. 299)
  a system of linear equations that has no solution

Example / Illustration

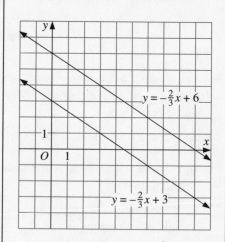

The graphs of $y = -\frac{2}{3}x + 6$
and $y = -\frac{2}{3}x + 3$ are parallel;
that is, no point lies on both
lines. The system

$$y = -\frac{2}{3}x + 6$$

$$y = -\frac{2}{3}x + 3$$

is an inconsistent system.

Dependent system (p. 299)
a system of linear equations that has infinitely many
solutions

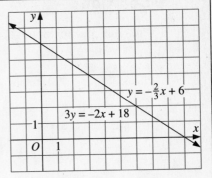

The graphs of $y = -\dfrac{2}{3}x + 6$ and
$3y = -2x + 18$ coincide; that
is, infinitely many points lie
on both lines. The system

$$y = -\frac{2}{3}x + 6$$

$$3y = -2x + 18$$

is a dependent system.

UNDERSTANDING THE MAIN IDEAS

Solving linear systems

By analyzing the coefficients of a system of linear equations, you can decide
how to solve the system. When the coefficients of one of the variables are the
same, you can use *subtraction* to solve the system algebraically. You can use
addition when the coefficients of one of the variables are *opposites* of
each other.

Example 1

Solve each system of equations.

  a. $4x - 3y = -15$ **b.** $2x - 2y = 12$
      $4x + 5y = -7$ $5x + 2y = 9$

■ Solution ■

a. The coefficients of the *x*-terms are the same. *Subtract* the second equation from the first.

$$4x - 3y = -15$$
$$4x + 5y = -7$$
$$\overline{ -8y = -8}$$ ← Divide both sides by –8.
$$y = 1$$

Now substitute $y = 1$ into either equation.

$$4x - 3y = -15$$
$$4x - 3(1) = -15$$
$$4x - 3 = -15$$
$$4x = -12$$
$$x = -3$$

The solution is $(x, y) = (-3, 1)$.

b. The coefficients of the *y*-terms are opposites. *Add* the two equations together.

$$2x - 2y = 12$$
$$5x + 2y = 9$$
$$\overline{7x = 21}$$ ← Divide both sides by 7.
$$x = 3$$

Substitute $x = 3$ into either equation.

$$2x - 2y = 12$$
$$2(3) - 2y = 12$$
$$6 - 2y = 12$$
$$-2y = 6$$
$$y = -3$$

The solution is $(x, y) = (3, -3)$.

Solve each system of equations by adding.

1. $3x - 2y = -8$
$5x + 2y = -8$

2. $6x - 2y = 9$
$-6x + 5y = -18$

3. $-5x + 3y = 10$
$-7x - 3y = 14$

Solve each system of equations by subtracting.

4. $2x + 8y = 30$
$5x + 8y = 39$

5. $10x + 2y = 3$
$10x - y = -9$

6. $3x + 2y = 9.5$
$-8x + 2y = -7$

If neither variable has coefficients that are the same or opposites, you can eliminate one of the variables by multiplying either one or both equations by a constant before adding or subtracting the resulting equations.

160

Example 2

Solve the system of equations $\begin{array}{l} 3x - 2y = 14 \\ 4x + 5y = -12 \end{array}$.

■ Solution ■

To eliminate the variable y, multiply the first equation by 5 and multiply the second equation by 2, and then add the resulting equations.

$$
\begin{array}{lll}
5(3x - 2y = 14) & \rightarrow & 15x - 10y = 70 \\
2(4x + 5y = -12) & \rightarrow & \underline{8x + 10y = -24} \\
& & 23x = 46 \\
& & x = 2
\end{array}
$$

Now substitute $x = 2$ into either *original* equation.

$$
\begin{array}{l}
3x - 2y = 14 \\
3(2) - 2y = 14 \\
6 - 2y = 14 \\
-2y = 8 \\
y = -4
\end{array}
$$

The solution is $(x, y) = (2, -4)$.

Solve each system of equations.

7. $2x + 5y = -14$
$ 5x + 3y = -16$

8. $3x + 2y = -3$
$ 2x - 7y = 23$

9. $4x - 3y = 1$
$ -8x + 7y = 3$

Inconsistent or dependent systems

Sometimes when solving a system of equations, both variables are eliminated and a mathematical statement results. If the statement is true, then there are infinitely many solutions of the system and the system is said to be *dependent*. If the statement is false, then there is no solution of the system and the system is said to be *inconsistent*. Graphically, the equations of a dependent system are the same line and the equations of an inconsistent system are two parallel lines.

Example 3

Solve each system of equations. Tell if the system is *inconsistent* or *dependent*.

a. $3x - 5y = 2$
$ 6x - 10y = 10$

b. $3x - 5y = 2$
$ 6x - 10y = 4$

■ Solution ■

a. Multiply the first equation by –2 and add.

$$-2(3x - 5y = 2) \qquad \rightarrow \qquad \begin{array}{r} -6x + 10y = -4 \\ 6x - 10y = 10 \\ \hline 0 = -6 \end{array}$$

The equation $0 = -6$ is never true. The system has *no solution*. It is an *inconsistent* system.

b. Multiply the first equation by –2 and add.

$$-2(3x - 5y = 2) \qquad \rightarrow \qquad \begin{array}{r} -6x + 10y = -4 \\ 6x - 10y = 4 \\ \hline 0 = 0 \end{array}$$

The equation $0 = 0$ is always true. The system has *infinitely many* solutions. It is a *dependent* system.

For Exercises 10–13, solve each system of equations. State whether the system has *one solution*, *infinitely many solutions*, or *no solution*.

10. $6x - 2y = -2$
   $3x - y = 5$

11. $10x - 4y = 9$
   $5x + 2y = 13$

12. $2x - 5y = 1$
   $5y - 2x = 0$

13. $4x - y = 3$
   $-8x + 2y = -6$

....
Spiral Review

Use substitution to solve each system of equations. *(Section 7.1)*

14. $y = 3x + 1$
   $2x + 2y = 5$

15. $y = 5 - 4x$
   $x = 5 - 4y$

For Exercises 16–18, use matrices *A* and *B* to evaluate each matrix expression. If an expression is not defined, write *undefined*. *(Section 1.3)*

$$A = \begin{bmatrix} 1 & 5 \\ -2 & 0 \end{bmatrix} \qquad B = \begin{bmatrix} 3 & 1 & 5 \\ 0 & 2 & 4 \end{bmatrix}$$

16. $5A$

17. $A - B$

18. $\frac{1}{2}B$

Solving Linear Systems with Matrices

Application

There is exactly one line through two distinct points, and there is exactly one parabola through three noncollinear points. You can use matrices to find the equation for such a line or parabola.

Learn how to . . .

- solve systems of linear equations using matrices

So you can . . .

- solve problems

Terms to Know	*Example / Illustration*
Identity matrices (p. 304) square matrices with 1's on the diagonal running from upper left to lower right, and 0's elsewhere	$\begin{bmatrix} 1 & 0 \\ 0 & 1 \end{bmatrix}$, $\begin{bmatrix} 1 & 0 & 0 \\ 0 & 1 & 0 \\ 0 & 0 & 1 \end{bmatrix}$, $\begin{bmatrix} 1 & 0 & 0 & 0 \\ 0 & 1 & 0 & 0 \\ 0 & 0 & 1 & 0 \\ 0 & 0 & 0 & 1 \end{bmatrix}$, and so on
Inverse matrix (p. 304) the square matrix that, when multiplied by a particular square matrix, gives an identity matrix (In symbols, the inverse of matrix A is written A^{-1}. *Note*: Not all square matrices have an inverse.)	$\begin{bmatrix} 0 & 3 \\ 2 & 1 \end{bmatrix} \cdot \begin{bmatrix} -\frac{1}{6} & \frac{1}{2} \\ \frac{1}{3} & 0 \end{bmatrix} = \begin{bmatrix} 1 & 0 \\ 0 & 1 \end{bmatrix}$ and $\begin{bmatrix} -\frac{1}{6} & \frac{1}{2} \\ \frac{1}{3} & 0 \end{bmatrix} \cdot \begin{bmatrix} 0 & 3 \\ 2 & 1 \end{bmatrix} = \begin{bmatrix} 1 & 0 \\ 0 & 1 \end{bmatrix}$ so both $\begin{bmatrix} 0 & 3 \\ 2 & 1 \end{bmatrix}$ and $\begin{bmatrix} -\frac{1}{6} & \frac{1}{2} \\ \frac{1}{3} & 0 \end{bmatrix}$ are inverses of each other. You can use a calculator or computer software with matrix capabilities to find inverse matrices.

Matrices and systems of equations

Matrices provide an abbreviated way to write a system of equations. The system can then be solved using matrix operations.

Example 1

Solve each system of equations using matrices.

a. $2x - 5y = 23$
$\quad\ x + 6y = -14$

b. $y = 3x - 1$
$\quad\ y = 3 + 5x$

c. $3x + y - 2z = 13$
$\quad\ x - 2y + 3z = -9$
$\quad -2x - 3y + z = -10$

■ Solution ■

Write each system in the form

$$\begin{bmatrix} \text{matrix of} \\ \text{coefficients} \end{bmatrix} \cdot \begin{bmatrix} x \\ y \end{bmatrix} = \begin{bmatrix} \text{matrix of} \\ \text{constants} \end{bmatrix}$$

(*Note*: Before writing the matrix equation, be sure that each equation is in the form $ax + by = c$.) Then solve the matrix equation by using a calculator or software with matrix calculation capabilities to multiply the inverse of the matrix of coefficients times the matrix of constants.

a. $\begin{bmatrix} 2 & -5 \\ 1 & 6 \end{bmatrix}\begin{bmatrix} x \\ y \end{bmatrix} = \begin{bmatrix} 23 \\ -14 \end{bmatrix}$

Then $\begin{bmatrix} x \\ y \end{bmatrix} = \begin{bmatrix} 2 & -5 \\ 1 & 6 \end{bmatrix}^{-1}\begin{bmatrix} 23 \\ -14 \end{bmatrix}$.

Using a calculator or software, $\begin{bmatrix} x \\ y \end{bmatrix} = \begin{bmatrix} 4 \\ -3 \end{bmatrix}$.

The solution is $(x, y) = (4, -3)$.

b. Rewrite each equation so it is in the form $ax + by = c$.

$y = 3x - 1 \quad\rightarrow\quad 3x - y = 1$
$y = 3 + 5x \quad\rightarrow\quad 5x - y = -3$

The matrix equation is $\begin{bmatrix} 3 & -1 \\ 5 & -1 \end{bmatrix}\begin{bmatrix} x \\ y \end{bmatrix} = \begin{bmatrix} 1 \\ -3 \end{bmatrix}$,

so $\begin{bmatrix} x \\ y \end{bmatrix} = \begin{bmatrix} 3 & -1 \\ 5 & -1 \end{bmatrix}^{-1}\begin{bmatrix} 1 \\ -3 \end{bmatrix} = \begin{bmatrix} -2 \\ -7 \end{bmatrix}$.

The solution is $(x, y) = (-2, -7)$.

(Solution continues on next page.)

---

c. The matrix equation is $\begin{bmatrix} 3 & 1 & -2 \\ 1 & -2 & 3 \\ -2 & -3 & 1 \end{bmatrix} \begin{bmatrix} x \\ y \\ z \end{bmatrix} = \begin{bmatrix} 13 \\ -9 \\ -10 \end{bmatrix}$.

So $\begin{bmatrix} x \\ y \\ z \end{bmatrix} = \begin{bmatrix} 3 & 1 & -2 \\ 1 & -2 & 3 \\ -2 & -3 & 1 \end{bmatrix}^{-1} \begin{bmatrix} 13 \\ -9 \\ -10 \end{bmatrix} = \begin{bmatrix} 2 \\ 1 \\ -3 \end{bmatrix}$.

The solution is $(x, y, z) = (2, 1, -3)$.

Use matrices to solve each system.

1. $3x - 4y = 2$
$x + 2y = 4$

2. $3.5x + y = -13.5$
$4.5x - 7y = -21.5$

3. $y = 5x + 7$
$y = -8x + 7$

4. $2x - 8 = 3y$
$5y = 2x - 20$

5. $2x - 3y + z = 17$
$x - 4y - 3z = -4$
$3x + y + z = 12$

6. $x + y + z = 4$
$3x - y - z = -2$
$5x + 3y + z = 5$

Matrices and equations for lines and parabolas

Matrices can be used to find an equation for the line through two distinct points or to find an equation for the parabola through three noncollinear points.

Example 2

Find an equation of:

a. the line passing through the points $(9, -4)$ and $(-3, -8)$.

b. the parabola passing through the points $(-2, 15)$, $(2, -1)$, and $(5, 8)$.

■ **Solution** ■

a. The equation of a line has the form $ax + b = y$. Substitute each ordered pair into this equation.

For $(9, -4)$: $9a + b = -4$
For $(-3, -8)$: $-3a + b = -8$

The matrix equation is $\begin{bmatrix} 9 & 1 \\ -3 & 1 \end{bmatrix} \begin{bmatrix} a \\ b \end{bmatrix} = \begin{bmatrix} -4 \\ -8 \end{bmatrix}$.

(*Solution continues on next page.*)

■ **Solution** ■ (*continued*)

Using a calculator or software with matrix calculation capabilities,

$$\begin{bmatrix} a \\ b \end{bmatrix} = \begin{bmatrix} 9 & 1 \\ -3 & 1 \end{bmatrix}^{-1} \begin{bmatrix} -4 \\ -8 \end{bmatrix} = \begin{bmatrix} \frac{1}{3} \\ -7 \end{bmatrix}.$$

So, $a = \frac{1}{3}$ and $b = -7$.

The equation of the line is $y = \frac{1}{3}x - 7$.

b. The equation of a parabola has the form $ax^2 + bx + c = y$. Substitute each ordered pair into this equation.

For $(-2, 15)$: $4a - 2b + c = 15$
For $(2, -1)$: $4a + 2b + c = -1$
For $(5, 8)$: $25a + 5b + c = 8$

The matrix equation is $\begin{bmatrix} 4 & -2 & 1 \\ 4 & 2 & 1 \\ 25 & 5 & 1 \end{bmatrix} \begin{bmatrix} a \\ b \\ c \end{bmatrix} = \begin{bmatrix} 15 \\ -1 \\ 8 \end{bmatrix}.$

Using a calculator or software with matrix calculation capabilities,

$$\begin{bmatrix} a \\ b \\ c \end{bmatrix} = \begin{bmatrix} 4 & -2 & 1 \\ 4 & 2 & 1 \\ 25 & 5 & 1 \end{bmatrix}^{-1} \begin{bmatrix} 15 \\ -1 \\ 8 \end{bmatrix} = \begin{bmatrix} 1 \\ -4 \\ 3 \end{bmatrix}.$$

So $a = 1$, $b = -4$, and $c = 3$.

The equation of the parabola is $x^2 - 4x + 3 = y$.

Find an equation of the line through each pair of points.

7. $\left(\frac{1}{2}, -\frac{11}{2} \right)$ and $\left(-\frac{3}{4}, \frac{3}{4} \right)$

8. $\left(-1, -\frac{10}{3} \right)$ and $\left(6, \frac{4}{3} \right)$

9. $(1, 3.5)$ and $(-4, -8)$

10. $(3, -12.6)$ and $(-5, 14.6)$

For Exercises 11 and 12, find an equation of the parabola through each set of points.

11. $(0, 5)$, $(2, -5)$, and $(4, -31)$

12. $(-4, 9)$, $(-1, 4.5)$, and $(3, 12.5)$

13. Mathematics Journal What do you suppose happens if you try to find an equation for a parabola through three *collinear* points? Choose three collinear points, and explain how you know they are collinear. Then use the procedure for finding the equation of a parabola described in Example 2. What are your results?

Study Guide, ALGEBRA 2: EXPLORATIONS AND APPLICATIONS

Solve each system of equations. State whether the system has *one solution*, *infinitely many solutions*, or *no solution*. *(Sections 7.1 and 7.2)*

14. $2x + 5y = 4$
$4x + 10y = 6$

15. $2x - 9y = 9$
$3x - 20y = 7$

16. $x - 3y = 5$
$6y = 2x - 10$

Solve each inequality for *x*. *(Toolbox, p. 787)*

17. $15x \le 3$

18. $6x - 5 > 25$

19. $3 - 9x < 1$

Write each quadratic function as a product of factors. *(Section 5.6)*

20. $y = 2x^2 - 5x - 3$

21. $y = x^2 - 12x + 35$

22. $y = x^2 - 3x - 4$

7.4

Inequalities in the Plane

Learn how to . . .

- write and graph
  inequalities with
  two variables

So you can . . .

- illustrate the
  relationship
  between two
  variable quantities

Application

As part of her training, an athlete runs at least 8 km per day. She can use an inequality to relate the total number of kilometers she runs to the number of days she trains.

UNDERSTANDING THE MAIN IDEAS

Graphs and inequalities

An inequality represents a region, called a *half-plane*, of the coordinate plane.

Example 1

Graph each inequality.

a. $y < \frac{2}{3}x + 6$

b. $y \geq -3x + 5$

c. $x \leq 7.5$

d. $-2x - 4y > 10$

■ Solution ■

a. Start by graphing the equation $y = \frac{2}{3}x + 6$ as a dashed line since it is not included in the graph of the inequality. Since y is *less than* $\frac{2}{3}x + 6$, the shaded half-plane *below* the dashed line is the graph of $y < \frac{2}{3}x + 6$.

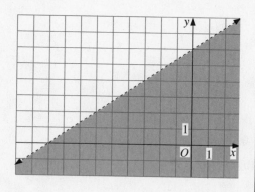

(Solution continues on next page.)

■ **Solution** ■ *(continued)*

b. Start by graphing the equation $y = -3x + 5$ as a solid line since it is included in the graph of the inequality. Since y is *greater than or equal to* $-3x + 5$, the line and the shaded half-plane *above* the line is the graph of $y \geq -3x + 5$.

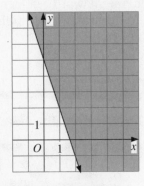

c. Begin by graphing the equation $x = 7.5$, which is a vertical line, as a solid line since it is included in the graph of the inequality. Since x is *less than or equal to* 7.5, the line and the shaded half-plane to the *left* of the line is the graph of $x \leq 7.5$.

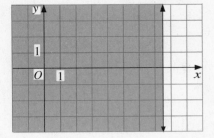

d. First, solve the inequality for y.

$$-2x - 4y > 10$$

$$-4y > 2x + 10$$

$$y < \frac{2}{-4}x + \frac{10}{-4}$$

$$y < -\frac{1}{2}x - \frac{5}{2}$$

Now graph the equation $y = -\frac{1}{2}x - \frac{5}{2}$ as a dashed line since it is not included in the graph of the inequality. Since y is *less than* $-\frac{1}{2}x - \frac{5}{2}$, the shaded half-plane *below* the dashed line is the graph of $y < -\frac{1}{2}x - \frac{5}{2}$ or $2x - 4y > 10$.

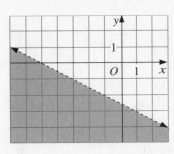

Graph each inequality.

1. $y \geq -x + 3$

2. $3x < 15$

3. $y \leq 4x + 1$

4. $x + y > 0$

5. $4x - 3y \leq 6$

6. $2x + 3 < -2$

7. $3y \geq y + 5$

8. $\frac{2}{3}y > -x + 1$

9. $2(x + y) > x + y$

A given half-plane can be described using an inequality. There are three determinations to be made regarding the graph. First, find an equation for the solid or dashed line shown. Then identify whether the shaded half-plane is above, below, to the left, or to the right of the line. Finally, identify whether the line is part of the graph.

Example 2

Find an inequality that defines each shaded region shown.

a. 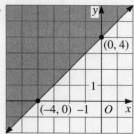 b. 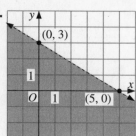 c.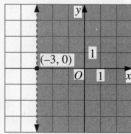

■ Solution ■

a. The line through $(-4, 0)$ and $(0, 4)$ has the equation

$$y = \left(\frac{4-0}{0-(-4)}\right)x + 4, \text{ or } y = x + 4.$$

The shaded half-plane is *above* the line, and the line is *solid* so it is part of the graph, so the correct inequality symbol is $\geq$. The inequality that defines the shaded region is $y \geq x + 4$.

b. The line through $(0, 3)$ and $(5, 0)$ has the equation

$$y = \left(\frac{0-3}{5-0}\right)x + 3, \text{ or } y = -\frac{3}{5}x + 3.$$

The shaded half-plane is *below* the line, and the line is *dashed* so it is not part of the graph, so the correct inequality symbol is $<$. The inequality that defines the shaded region is $y < -\frac{3}{5}x + 3$.

c. A vertical line has the form $x = k$, so an equation for the line is $x = -3$. The shaded half-plane is to the *right* of the line, and the line is *dashed* so it is not part of the graph, so the correct inequality symbol is $>$. The inequality that defines the shaded region is $x > -3$.

Find an inequality that defines each shaded region.

10.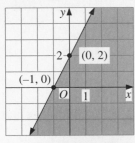

11.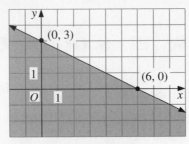

12.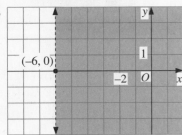

13.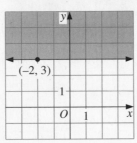

An athlete trains by running at least 8 km per day.

14. How far does she run in 5 days? 10 days? 30 days?

15. Suppose the athlete has run exactly 100 km since she began training. Is it possible that she has been training for 14 days? Explain your answer.

16. Using a coordinate plane with "Number of days training" on the *x*-axis and "Distance run (km)" on the *y*-axis, draw a graph to model this relationship.

· · · · · · · · · · · · · · · · · · ·

Spiral Review

For Exercises 17–19, solve each system of equations. *(Section 7.2)*

17. $2x - 4y = 8$
$3x + y = -9$

18. $2x - 4y = 4$
$3x + 2y = 10$

19. $7x - 3y = 20$
$8x + 5y = 65$

20. An athlete in training records the following distance-time pairs for twelve days of training. *(Section 2.4)*

Day	Distance (km)	Time (min)	Day	Distance (km)	Time (min)
1	8	25.6	7	10	35.0
2	9	32.4	8	11	35.2
3	8	27.2	9	12	40.8
4	10	32.0	10	10	34.0
5	10	33.0	11	12	42.0
6	12	40.8	12	15	49.5

a. Record the data as distance-time ordered pairs and plot them in a coordinate plane.

b. Using a calculator or software with statistics capabilities, find the equation of a line that models the ordered pairs.

c. What does the slope of the line you found in part (b) represent?

Systems of Inequalities

Learn how to . . .

- graph a system of linear inequalities
- find the system of linear inequalities that defines a given graph

So you can . . .

- represent and interpret situations involving inequalities

Application

The graph below models the relationship between a student's study times and his test scores for 15 tests.

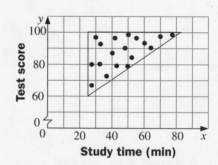

For this data, all the points can be enclosed in a triangular region. The student can use a system of inequalities to describe the relationship between study time and test score.

Terms to Know

System of inequalities (p. 315)
a set of inequalities that describes a region of a coordinate plane

Example / Illustration

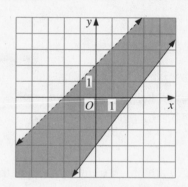

Since the dashed line is *not* part of the shaded region and the solid line *is* part of it, the system of inequalities describing the shaded region is

$$y < x + 2$$
$$y \geq \frac{4}{3}x - 3.$$

Systems of inequalities and regions of a coordinate plane

You can model a system of inequalities by shading the appropriate region of a coordinate plane.

Example 1

Graph the system of inequalities $\begin{array}{l} y \le x + 5 \\ x < 6 \end{array}$.

■ Solution ■

The top figure at the right shows the graphs of $y = x + 5$ and $x = 6$. The half-plane for $y \le x + 5$ is the line $y = x + 5$ and the region below the line. The half-plane for $x < 6$ is just the region to the left of the line $x = 6$. (It does not include the line $x = 6$.) The graph of the system of inequalities is the region where these two half-planes overlap. The bottom figure at the right shows the graph of the system. Notice that the line $x = 6$ is shown as a dashed line.

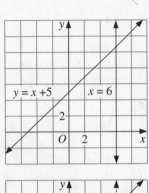

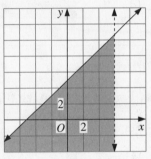

$y \le x + 5$
$x < 6$

The procedure for graphing a system of more than two inequalities is similar to that shown in Example 1. The half-plane for each inequality is determined, and the region (if one exists) where *all* of these half-planes overlap is the graph of the system of inequalities.

For Exercises 1–6, graph each system of inequalities.

1. $y \ge 2x - 5$
$y < 2x + 3$

2. $y > \dfrac{2}{3}x - 2$
$y \le -x + 5$

---

3. $y \geq x + 2$
$\quad y < \frac{1}{2}x + 6$
$\quad x \geq -1$

4. $y < 5$
$\quad y > 2$
$\quad x > 3$
$\quad x < 6$

5. $x \geq 0$
$\quad y \geq 0$

6. $x \geq 6$
$\quad x \leq -3$

You can write a system of inequalities to represent a shaded region.

Example 2

Find a system of inequalities defining each shaded region.

a.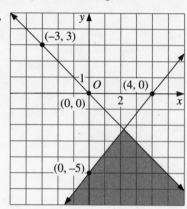

b.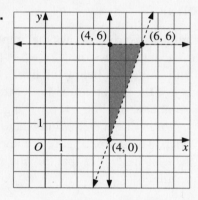

▪ Solution ▪

a. For the line through $(0, -5)$ and $(4, 0)$, the slope is $\frac{0 - (-5)}{4 - 0} = \frac{5}{4}$, so an equation for this line is $y = \frac{5}{4}x - 5$. For the line through $(-3, 3)$ and $(0, 0)$, the slope is $\frac{0 - 3}{0 - (-3)} = -1$, so an equation for this line is $y = -x$. The shaded region is below each line and each line is solid, so the system is

$$y \leq \frac{5}{4}x - 5$$
$$y \leq -x$$

b. An equation for the line through $(4, 0)$ and $(6, 6)$ is $y = 3x - 12$. The shaded region is above this line. An equation for the line through $(4, 6)$ and $(6, 6)$ is $y = 6$. The shaded region is below this line. An equation for the line through $(4, 0)$ and $(4, 6)$ is $x = 4$. The shaded region is to the right of this line. Also, the vertical line is solid and the other two lines are dashed. So the system of inequalities is

$$y > 3x - 12$$
$$y < 6$$
$$x \geq 4$$

For Exercises 7–10, write a system of inequalities defining each shaded region.

7.
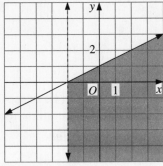
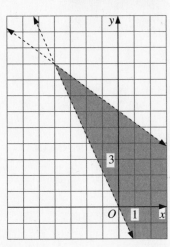

8.
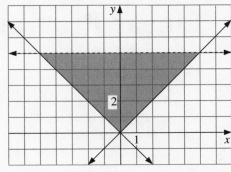
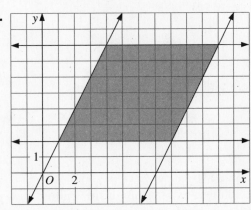

9.

10.

11. Write a system of inequalities that defines the region between the lines $3x - 4y = -28$ and $-x + 2y = -8$, including both lines.

12. Write a system of inequalities that defines the interior of the triangle with vertices $(0, 5)$, $(-10, 4)$, and $(0, 0)$.

· · · · · · · · · · · · · · · · · · · ·
Spiral Review

Graph each inequality. *(Section 7.4)*

13. $3x - 2y \le 1$

14. $-3y > 16$

15. $3x + y < 2$

Solve each system of equations. *(Section 7.1)*

16. $3x + y = 3$
    $-4x - y = -5$

17. $x = 8y$
    $2.5x - 0.5y = -19.5$

18. $4x + 6y = 22$
    $3y = 1$

Find the *x*-intercepts of the graph of each equation. *(Section 5.3)*

19. $y = (2x + 1)(x - 3)$

20. $y = (-5x + 1)(4x - 9)$

21. $y = 5(2x + 7)(2x - 7)$

Linear Programming

Learn how to . . .

- find the best solution when several conditions have to be met

So you can . . .

- solve problems involving several linear inequalities

Application

For a maker of compact disks and laser disks, production costs can be described as a system of inequalities. The manufacturer can use linear programming to find the number of compact disks and the number of laser disks that minimize costs and maximize profit.

Terms to Know

Example / Illustration

Feasible region (p. 323)

a graph of the solution of a system of inequalities in a linear programming problem (The region represents all the given constraints.)

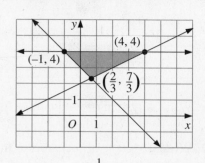

$$y \geq \frac{1}{2}x + 2$$
$$y \geq -x + 3$$
$$y \leq 4$$

The point $(1, 3)$ is in the feasible region. The point $(1, 6)$ is not in the feasible region.

Corner-point principle (p. 323)

the principle which states that for a polygonal feasible region the best solution will be at a corner point of the feasible region

The corner points of the feasible region above are $(-1, 4)$, $(4, 4)$, and $\left(\frac{2}{3}, \frac{7}{3}\right)$.

One of these three points represents the best solution to the linear programming problem.

Study Guide, ALGEBRA 2: EXPLORATIONS AND APPLICATIONS

Linear programming (p. 323) the method of finding the best solution for a polygonal feasible region	If the profit function $P = 20x + 30y$ is to be maximized under the constraints that defined the feasible region shown on the previous page, then the maximum profit, 200, occurs at the corner-point (4, 4).

UNDERSTANDING THE MAIN IDEAS

If you have a feasible region and a profit equation, you can use linear programming to find the point in the feasible region that yields the maximum value of the profit equation.

Example

In the manufacturing of compact disks and laser disks, a company's expenses are described by the following system of inequalities.

$$x \geq 0$$
$$y \geq 0$$
$$y \leq \frac{3}{2}x + 12$$
$$y \leq -10x + 150$$

The feasible region for this system of inequalities is shown at the right.

The company's profits are given by the equation $P = 7x + 5y$. Find the company's maximum profit.

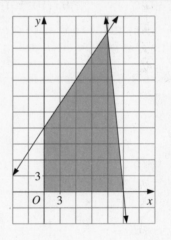

■ Solution ■

Step 1: Find the coordinates of the four corner points of the feasible region.

One corner point is the origin, or (0, 0). A second corner point is the y-intercept of the line $y = \frac{3}{2}x + 12$, or (0, 12). A third corner point is the x-intercept of the graph of the line $y = -10x + 150$, or (15, 0). The fourth corner point is the point of intersection of the lines $y = \frac{3}{2}x + 12$ and $y = -10x + 150$; using substitution to solve the equations simultaneously, this point is (12, 30).

(Solution continues on next page.)

■ **Solution** ■ *(continued)*

Step 2: Substitute the coordinates of each ordered pair into the profit equation.

Evaluate $P = 7x + 5y$ at each corner point.

Corner point	Value of P
(0, 0)	$7(0) + 5(0) = 0$
(0, 12)	$7(0) + 5(12) = 60$
(12, 30)	$7(12) + 5(30) = 234$
(15, 0)	$7(15) + 5(0) = 105$

The profit is maximized when $x = 12$ and $y = 15$.

1. The following points lie in the feasible region discussed in the Example. Show that each point does *not* give the maximum value for P.

   a. (0, 10)   **b.** (10, 10)   **c.** (11, 30)   **d.** (14, 10)

2. Explain why it does not make sense to use the point (20, 25) in the profit equation $P = 7x + 5y$ discussed in the Example.

For Exercises 3–6, graph each feasible region. Label the corner points of each region.

3. $x \geq 0$
   $y \geq 0$
   $y \leq -\frac{1}{2}x + 6$

4. $x \leq 5$
   $y \geq -1$
   $y \leq \frac{2}{3}x + \frac{5}{3}$

5. $y \leq \frac{4}{3}x + 4$
   $y \leq -4x + 20$
   $y \geq 0$

6. $y \leq x + 4$
   $y \leq 4$
   $y \geq 0$
   $y \leq -x + 8$

7. For each of the feasible regions in Exercises 3–6, evaluate the profit function $P = 7x + 8y$ for each corner point. Identify the corner point of each feasible region that maximizes this profit function.

8. **Mathematics Journal** Describe how the skills of solving systems of equations and graphing inequalities relate to the skills of using the corner-point principle and linear programming.

For each function, find the inverse. *(Section 4.1)*

9. $f(x) = 3x - 5$

10. $g(x) = 1 - \dfrac{1}{2}x$

11. $h(x) = x$

12. $f(x) = \dfrac{4x - 3}{2}$

Tell why each question might be biased. Then describe what changes you would make to improve the question. *(Section 6.2)*

13. "Should it be easier for students to park near the high school?"

14. "How much money do you spend on clothes each year?"

Chapter 7 Review ··············

Complete these exercises for a review of Chapter 7. If you have difficulty with a particular problem, review the indicated section.

Use substitution to solve each system of equations. *(Section 7.1)*

1. $4x + y = -1$
$x = \frac{1}{2}y + 5$

2. $y = 2x + 1$
$4x - 2y = -2$

TECHNOLOGY Use a graphing calculator or graphing software to solve each system of equations. (The graphs of the equations may intersect more than once.) *(Section 7.1)*

3. $y = x^2 + 2x + 1$
$y = -x + 2$

4. $y = x^2 - 4$
$y = -x^2 + 1$

For Exercises 5–8, solve each system of equations. *(Section 7.2)*

5. $4x - 3y = 17$
$5x - 3y = 19$

6. $2x - 5y = 29$
$3x - 2y = 5$

7. $-15x + 2y = -9$
$-3x - 2y = -3$

8. $-8x - 5y = -7$
$8x + 5y = 8$

9. What value of a will make the following system of linear equations inconsistent? *(Section 7.2)*

$$3x - 2y = 5$$
$$3x + ay = -5$$

10. What values of b and c will make the following system of linear equations dependent? *(Section 7.2)*

$$5x + 3y = -4$$
$$10x + by = c$$

11. Write the system of equations that is represented by the matrix equation

$\begin{bmatrix} 3 & 1 \\ 5 & 2 \end{bmatrix} \begin{bmatrix} x \\ y \end{bmatrix} = \begin{bmatrix} -2.4 \\ -3.6 \end{bmatrix}$. *(Section 7.3)*

12. Use a calculator or software with matrix calculation capabilities to solve the matrix equation in Exercise 11 for x and y. *(Section 7.3)*

13. Writing Explain how you can use a calculator or software with matrix calculation capabilities to solve a system of linear equations. *(Section 7.3)*

Graph each inequality. *(Section 7.4)*

14. $y \geq -4x$

15. $x \leq 3$

16. $y > \frac{1}{2}x - 3$

17. $y \leq -4x - 4$

Graph each system of inequalities. *(Section 7.5)*

18. $y \le \frac{3}{4}x + 3$

$y > -\frac{3}{4}x$

19. $2x + 3y \le 6$

$2x - 3y \le 15$

For Exercises 20 and 21, write a system of inequalities defining each shaded region. *(Section 7.5)*

20.

21.

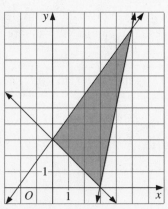

22. Open-ended On a coordinate plane, select four noncollinear ordered pairs as the vertices of a quadrilateral. Then find a system of inequalities that defines the interior of the quadrilateral. *(Section 7.5)*

For Exercises 23 and 24, use the shaded region given in Exercise 21 as a feasible region, and use the equation $P = 12x + 3y$ as a profit function. *(Section 7.6)*

23. Find the value of P at each corner point of the feasible region.

24. Which corner point maximizes the profit function?

SPIRAL REVIEW   Chapters 1–7

Use matrices $A = \begin{bmatrix} 6 & 4 \\ -2 & 3 \end{bmatrix}$ and $B = \begin{bmatrix} 1 & 4 \\ 0 & 2 \end{bmatrix}$ to find each of the following.

1. $A + B$

2. $3A - B$

3. $-2(AB)$

Evaluate each expression.

4. $125^{2/3}$

5. $\frac{1}{64^{3/2}}$

6. $\frac{25^0}{25^{1/2}}$

7. $\log_3 9$

8. $\log_5 1$

9. $3x^2 - 7$ for $x = \frac{1}{3}$

Solve for x.

10. $\log_x 125 = \frac{3}{2}$

11. $\log_{100} x = \frac{1}{4}$

12. $6x^2 = 96$

13. $\frac{x^2}{5} = 4$

14. $x^2 - 3x + 2 = 0$

15. $x^2 + 5x - 2 = 0$

Use the function $f(x) = 2x^2 - 4x + 7$.

16. Rewrite the function in the form $f(x) = a(x - h)^2 + k$.

17. What is the minimum value of the function?

Study Guide, ALGEBRA 1: EXPLORATIONS AND APPLICATIONS

Square Root Functions

Learn how to . . .

- restrict the domain of a function to obtain an inverse function

- graph and evaluate square root functions

So you can . . .

- understand the dynamics of situations described by square root functions

Application

When a wave travels in the water, its speed changes as the depth of the water changes. You can use the square root function

$$s = \sqrt{35.28d},$$

where s is the speed (in km/s) and d is the depth (in km), to relate wave speed and water depth.

UNDERSTANDING THE MAIN IDEAS

Finding an inverse function

When you take the inverse of a quadratic function, such as $f(x) = x^2$, the result is a radical function. Recall that one way to find the inverse of a function is to substutute y for $f(x)$, switch the variables x and y, and then solve for y. For the function $f(x) = x^2$, the result of this method is $y = \pm\sqrt{x}$ which is not a function since for each positive value of x there are two values of y. In order for the inverse of $f(x) = x^2$ to be a function, we must place a restriction on the domain of $f(x) = x^2$. There are two ways to do this: let $f(x) = x^2$ for $x \geq 0$ and let $f(x) = x^2$ for $x \leq 0$. Then the inverse of $f(x) = x^2$ for $x \geq 0$ is $f^{-1}(x) = \sqrt{x}$ and the inverse of $f(x) = x^2$ for $x \leq 0$ is $f^{-1}(x) = -\sqrt{x}$.

Example 1

For the function $f(x) = \frac{1}{4}x^2$, describe the inverse function using two equations.

■ Solution ■

$$y = \frac{1}{4}x^2 \qquad \leftarrow \text{Substitute } y \text{ for } f(x).$$

$$\pm\sqrt{y} = \frac{1}{2}x \qquad \leftarrow \text{Take the square root of both sides.}$$

$$x = \pm 2\sqrt{y} \qquad \leftarrow \text{Solve for } x.$$

$$y = \pm 2\sqrt{x} \qquad \leftarrow \text{Switch } x \text{ and } y.$$

The equation $y = \pm 2\sqrt{x}$ does not represent a function because each possible x-value except zero has two y-values. The domain of the original function

$y = \frac{1}{4}x^2$ must be restricted in order to obtain the inverse function.

If the domain of $y = \frac{1}{4}x^2$ is $x \geq 0$, then the inverse is $y = 2\sqrt{x}$.

If the domain of $y = \frac{1}{4}x^2$ is $x \leq 0$, then the inverse is $y = -2\sqrt{x}$.

In Example 1, notice that the *range* of the original function, $f(x) = \frac{1}{4}x^2$, is $f(x) \geq 0$. This corresponds to the fact that the domain of each inverse function is $x \geq 0$.

For each function, restrict the domain of the original function and write the inverse function as two equations.

1. $y = x^2$ 　　　　　 **2.** $y = \frac{1}{16}x^2$ 　　　　　 **3.** $f(x) = 9x^2$

4. $f(x) = 100x^2$ 　　　 **5.** $y = x^2 + 1$ 　　　　　 **6.** $y = 4x^2 - 2$

Graphing square root functions

Just as the graph of a function $y = a(x - h)^2 + k$ is a translation of the graph of $y = ax^2$, the graph of a function such as $y = \sqrt{x - 3} + 2$ is a translation of the graph of $y = \sqrt{x}$.

Example 2

Compare the graphs of $y = \sqrt{x - 3} + 2$ and $y = \sqrt{x}$. The domain of the function $y = \sqrt{x}$ is $x \geq 0$ and the range is $y \geq 0$. Use your graph to state the domain and range of the function $y = \sqrt{x - 3} + 2$.

■ Solution ■

Using a graphing calculator, graph the two functions on the same screen.
The graphs are shown below.

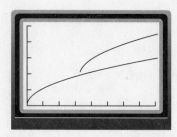

The graphs have the same shape. The graph of $y = \sqrt{x-3} + 2$ is a translation 3 units
to the right and 2 units up of the graph of $y = \sqrt{x}$. The domain of the function
$y = \sqrt{x-3} + 2$ is $x \geq 3$ and the range of the function is $y \geq 2$.

For Exercises 7–10, graph each function. State the domain and range.

7. $y = \sqrt{x} + 2$

8. $y = \sqrt{x-2}$

9. $y = \sqrt{x+2}$

10. $y = -(\sqrt{x} + 2)$

11. Refer to the Application at the beginning of this section. What is the average
speed of a wave passing through water that is 30 km deep?

······················
Spiral Review

12. The set of inequalities below model the constraints on the manufacturing
process for a certain company. *(Section 7.6)*

$$x + y \leq 40$$
$$x + y \geq 10$$
$$x \geq 5$$
$$y \geq 0$$

a. Graph the feasible region for these inequalities. Label the coordinates of
the vertices of the region.

b. Evaluate each corner point of the feasible region for the profit function
$P = 2x + 3y$.

c. Which point maximizes the profit function?

Evaluate each expression. *(Section 3.2)*

13. $64^{1/2}$

14. $64^{1/6}$

15. $64^{1/3}$

Graph each inequality. *(Section 7.4)*

16. $y \leq x$

17. $3x - 2y \leq -6$

18. $-2x + 4y < 0$

Radical Functions

Application

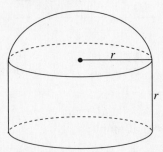

The package for a brand of nylon stockings is a hemisphere on top of a cylinder, where the radius of the cylinder (and hemisphere) is the same as the height of the cylinder. The radius r of the hemisphere can be expressed as a radical function of the volume V of the entire package.

Terms to Know	Example / Illustration

Radical function (p. 344)
 a function of the form $y = \sqrt[n]{x}$, where the domain of x is restricted so that y is a real number (A radical function has a domain and range of all numbers when n is odd, and a domain and range of nonnegative numbers when n is even.)
	$y = \sqrt[3]{x}$
	   domain: all real numbers x
	   range: all real numbers y
	
	$y = \sqrt[4]{x}$ domain: $x \geq 0$
	   range: $y \geq 0$
	
	Notice that you can find an odd root of a positive or negative number, but you can only find even roots of positive numbers.

UNDERSTANDING THE MAIN IDEAS

The meaning of $\sqrt[n]{b}$

The symbol $\sqrt{x}$, which is read "the square root of x," can also be written as $\sqrt[2]{x}$.

The expressions $\sqrt[3]{-8}$ and $\sqrt[4]{z^8}$ are read as "the cube root of -8" and "the fourth root of the eighth power of z," respectively.

Example 1

State the domain and range of each function.

a. $y = \sqrt[4]{x}$ **b.** $y = \sqrt[3]{x}$ **c.** $f(x) = \sqrt[6]{x} + 7$

d. the inverse of $f(x) = \frac{1}{64}x^6$, for $x \le 0$

Solution

a. For an even root to be defined, x must be nonnegative. So, the domain of $y = \sqrt[4]{x}$ is $x \ge 0$. The range of the function is $y \ge 0$.

b. You can find an odd root of any number. So, the domain of $y = \sqrt[3]{x}$ is all numbers. The range of the function is also all numbers.

c. Since the value of n is even, the domain of $f(x) = \sqrt[6]{x} + 7$ is $x \ge 0$. Since $\sqrt[6]{x} \ge 0$ for $x \ge 0$, the range of the function is $f(x) \ge 7$.

d. If $x \le 0$ for the original function, the inverse function is $y = -\sqrt[6]{64x} = -2\sqrt[6]{x}$. For this function, the domain is $x \ge 0$ and the range is $y \le 0$.

State the domain and range for each function.

1. $y = \sqrt[3]{2x}$ **2.** $f(x) = -\sqrt[5]{x}$ **3.** $y = \sqrt[6]{5x}$

4. $g(x) = \sqrt[4]{-x}$ **5.** $y = \sqrt[4]{x} - 4$ **6.** the inverse of $g(x) = \frac{1}{25}x^3$, $x \le 0$

Extending the properties of exponents to radical expressions

You can extend the properties of exponents to radicals as long as the value of a and b below are not both negative.

Power of Product: $(ab)^m = a^m b^m$ $\rightarrow$ $\sqrt[m]{ab} = \sqrt[m]{a} \cdot \sqrt[m]{b}$

Power of a Quotient: $\left(\dfrac{a}{b}\right)^m = \dfrac{a^m}{b^m}$ $\rightarrow$ $\sqrt[m]{\dfrac{a}{b}} = \dfrac{\sqrt[m]{a}}{\sqrt[m]{b}}$

Power of a Power: $(b^m)^n = b^{mn} = b^{nm} = (b^n)^m$ $\rightarrow$ $\sqrt[m]{b^n} = (\sqrt[m]{b})^n$

Example 2

Convert between fractional exponents and radical notation.

a. $100^{5/2}$ **b.** $\sqrt[5]{s^{20}}$ **c.** $\sqrt[3]{-\dfrac{m^{15}}{64}}$ **d.** $(3y)^{2/3}$

■ **Solution** ■

a. $100^{5/2} = (100^{1/2})^5$
$\qquad\qquad = (\sqrt{100})^5 \qquad \leftarrow$ by definition, $a^{1/n} = \sqrt[n]{a}$
$\qquad\qquad = 10^5$
$\qquad\qquad = 100{,}000$

b. $\sqrt[5]{s^{20}} = (s^{20})^{1/5} \qquad \leftarrow$ by definition, $\sqrt[n]{a} = a^{1/n}$
$\qquad\quad\; = s^{20/5}$
$\qquad\quad\; = s^4$

c. $\sqrt[3]{-\dfrac{m^5}{64}} = \sqrt[3]{-\dfrac{1}{64} \cdot m^5} \qquad \leftarrow$ power of a product

$\qquad\qquad = \sqrt[3]{-\dfrac{1}{64}} \cdot \sqrt[3]{m^5}$

$\qquad\qquad = \dfrac{\sqrt[3]{-1}}{\sqrt[3]{64}} \cdot \sqrt[3]{m^5} \qquad \leftarrow$ power of a quotient

$\qquad\qquad = \dfrac{-1}{4} \cdot m^{5/3}$

$\qquad\qquad = -\dfrac{m^{5/3}}{4}$

d. $(3y)^{2/3} = (3y)^{2 \,\cdot\, 1/3}$
$\qquad\qquad = ((3y)^2)^{1/3}$
$\qquad\qquad = \sqrt[3]{(3y)^2}$
$\qquad\qquad = \sqrt[3]{9y^2}$

Express using fractional exponents.

7. $\sqrt[4]{m}$ **8.** $\sqrt{25s}$ **9.** $\sqrt[3]{n^5}$ **10.** $\sqrt[5]{m^{30}}$

Express using radical notation.

11. $p^{1/5}$ **12.** $r^{7/3}$ **13.** $(-3s)^{1/3}$ **14.** $y^{m/3}$

Simplifying radical expressions

The properties of exponents can be used to simplify radical expressions.

Example 3

Simplify each expression. Assume $p \geq 0$ in part (b) and $t \geq 0$ in part (c).

a. $\sqrt[3]{-16}$ **b.** $\sqrt[4]{5p^{12}}$ **c.** $\sqrt[5]{\dfrac{1}{32}t^{13}}$

Solution

a. $\sqrt[3]{-16} = \sqrt[3]{(-8)(2)}$

$\quad\quad\quad = \sqrt[3]{-8} \cdot \sqrt[3]{2} \quad \leftarrow \sqrt[3]{-8} = -2$ because $(-2)^3 = -8$

$\quad\quad\quad = -2\sqrt[3]{2}$

b. $\sqrt[4]{5p^{12}} = \sqrt[4]{5} \cdot \sqrt[4]{p^{12}}$

$\quad\quad\quad = \sqrt[4]{5} \cdot (p^{12})^{1/4}$

$\quad\quad\quad = \sqrt[4]{5} \cdot p^{12/4}$

$\quad\quad\quad = p^3 \sqrt[4]{5}$

c. $\sqrt[5]{\dfrac{1}{32}t^{13}} = \sqrt[5]{\dfrac{1}{32}} \cdot \sqrt[5]{t^{13}}$

$\quad\quad\quad = \sqrt[5]{\dfrac{1}{32}} \cdot \sqrt[5]{t^{10+3}}$

$\quad\quad\quad = \sqrt[5]{\dfrac{1}{32}} \cdot \sqrt[5]{t^{10}t^3} \quad\quad \leftarrow a^{m+n} = a^m a^n$

$\quad\quad\quad = \sqrt[5]{\dfrac{1}{32}} \cdot \sqrt[5]{t^{10}} \cdot \sqrt[5]{t^3} \quad \leftarrow \sqrt[m]{ab} = \sqrt[m]{a} \cdot \sqrt[m]{b}$

$\quad\quad\quad = \dfrac{1}{2}t^2 \sqrt[5]{t^3}$

Simplify each radical expression. Assume $m \geq 0$ in Exercise 16.

15. $\sqrt[3]{-512}$ **16.** $\sqrt[4]{10m^{36}}$

Use fractional exponents and properties of exponents to prove each statement. Assume $x \geq 0$.

17. $\dfrac{\sqrt[5]{x^4}}{\sqrt[10]{x^3}} = \sqrt{x}$ **18.** $(\sqrt[n]{x})^n + \sqrt[n]{x^n} = 2x$

For Exercises 19–21, refer to the Application at the beginning of this section.

19. Express the volume V as a function of r, the radius and height of the cylinder. (Use the formula $V = \pi r^2 h$ for the volume of a cylinder and the formula $V = \dfrac{2}{3}\pi r^3$ for the volume of a hemisphere.)

20. Write an equation for r as a function of V.

21. Suppose V is 1200 cm^3. Find r.

22. Mathematics Journal Which seems easier to work with, fractional exponents or radical notation? Which one is simpler to enter into a calculator?

For Exercises 23 and 24, graph each function. State the domain and range.
(Section 8.1)

23. $f(x) = \sqrt{x} - 5$

24. $g(x) = -\sqrt{\dfrac{x}{2}}$

25. An object falls according to the equation $d = 16t^2$, where d is the distance in feet and t is the time in seconds. *(Section 5.1)*

   a. How far does an object fall during the third second?

   b. How long does it take an object to fall 500 ft?

Solve. Check to eliminate any extraneous solutions. *(Section 4.4 and 5.6)*

26. $\log_3 (x + 3) - \log_3 (x - 1) = 2$

27. $\log_{12} (x - 11) + \log_{12} x = 1$

Section 8.3 Solving Radical Equations

Learn how to . . .

- solve equations with radical expressions

So you can . . .

- solve problems that can be modeled using radical expressions

Application

Outside on a winter day, the wind chill makes it feels colder if a strong wind is blowing. You can use a radical equation to describe how cold it feels on a windy day.

UNDERSTANDING THE MAIN IDEAS

Solving radical equations graphically

You can use a graphing calculator or graphing software to solve a radical equation.

Example 1

Use a graphing calculator or graphing software to solve the radical equation

$$\frac{1}{4}\sqrt{d} + \frac{1}{1116}d = 12.$$

■ Solution ■

To solve the equation graphically, graph the related equations $y = \frac{1}{4}\sqrt{x} + \frac{1}{1116}x$
(the left side of the radical equation) and $y = 12$ (the right side of the radical equation) on the same coordinate axes. The figure below shows the graphs.

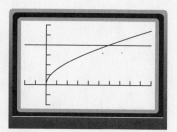

(Solution continues on next page.)

The *x*-coordinate of the point where these two graphs intersect represents the value of *d* that is the solution of given the radical equation. Using the Trace feature, this *x*-coordinate is about 1745. So the solution is $d \approx 1745$.

TECHNOLOGY Use a graphing calculator or graphing software to solve each radical equation.

1. $\sqrt{x} + \sqrt[4]{x} = 5$ **2.** $3\sqrt{x} + 2\sqrt[3]{x} = 10$

Solving radical equations algebraically

You can solve a radical equation algebraically. If necessary, isolate the radical expression on one side of the equation. Then raise both sides of the equation to the power that will eliminate the radical. Solve the resulting equation for the variable. This method sometimes produces extraneous solutions, so all possible solutions must be checked in the original equation.

Example 2

Solve each radical equation.

a. $\sqrt{x + 5} = 7$ **b.** $\sqrt{4x + 24} = x + 3$

c. $\sqrt[5]{y - 3} = 2$ **d.** $\sqrt{2x + 3} = x - 1$

■ **Solution** ■

a. $\sqrt{x + 5} = 7$

$\qquad (\sqrt{x + 5})^2 = 7^2$ ← Square both sides; $(\sqrt{a})^2 = a$.

$\qquad\qquad x + 5 = 49$

$\qquad\qquad\qquad x = 44$

Check the solution: $\sqrt{44 + 5} \overset{?}{=} 7$

$\qquad\qquad\qquad\qquad\qquad \sqrt{49} \overset{?}{=} 7$

$\qquad\qquad\qquad\qquad\qquad\qquad 7 = 7 \checkmark$

The solution of the equation $\sqrt{x + 5} = 7$ is 44.

(Solution continues on next page.)

■ Solution ■ *(continued)*

b. $\sqrt{4x + 24} = x + 3$

$4x + 24 = x^2 + 6x + 9 \quad \leftarrow$ Square both sides.

$0 = x^2 + 2x - 15$

$0 = (x + 5)(x - 3)$

$x = -5 \text{ or } x = 3$

Check each solution:

$\sqrt{4(-5) + 24} \stackrel{?}{=} -5 + 3 \qquad \sqrt{4(3) + 24} \stackrel{?}{=} 3 + 3$

$\sqrt{4} \stackrel{?}{=} -2 \qquad\qquad\qquad \sqrt{36} \stackrel{?}{=} 6$

$2 \neq -2 \qquad\qquad\qquad\qquad 6 = 6 \checkmark$

The solution $x = -5$ does not check, it is extraneous.
The solution of the equation $\sqrt{4x + 24} = x + 3$ is 3.

c. $\sqrt[5]{y - 3} = 2$

$y - 3 = 2^5 \quad \leftarrow$ Raise both sides to the fifth power.

$y - 3 = 32$

$y = 35$

Check the solution: $\quad \sqrt[5]{35 - 3} \stackrel{?}{=} 2$

$\sqrt[5]{32} \stackrel{?}{=} 2$

$2 = 2 \checkmark$

The solution of the equation $\sqrt[5]{y - 3} = 2$ is 2.

d. $\sqrt{2x + 3} = x - 1$

$2x + 3 = x^2 - 2x + 1$

$0 = x^2 - 4x - 2$

Use the quadratic formula to solve this equation.

$x = \dfrac{-b \pm \sqrt{b^2 - 4ac}}{2a}$

$x = \dfrac{-(-4) \pm \sqrt{(-4)^2 - 4(1)(-2)}}{2(1)} \quad \leftarrow a = 1, b = -4, c = -2$

$= \dfrac{4 \pm \sqrt{16 + 8}}{2}$

$= \dfrac{4 \pm \sqrt{24}}{2}$

$= 2 \pm \sqrt{6}$

So, $x = 2 + \sqrt{6} \approx 4.45 \text{ or } x = 2 - \sqrt{6} \approx -0.45.$

(Solution continues on next page.)

Check each solution:

$$\sqrt{2x + 3} = x - 1 \qquad\qquad \sqrt{2x + 3} = x - 1$$

$$\sqrt{2(4.45) + 3} \stackrel{?}{=} 4.45 - 1 \qquad \sqrt{2(-0.45) + 3} \stackrel{?}{=} -0.45 - 1$$

$$\sqrt{8.9 + 3} \stackrel{?}{=} 3.45 \qquad\qquad \sqrt{-0.9 + 3} \neq -1.45$$

$$\sqrt{11.9} \approx 3.45 \checkmark \qquad\qquad \text{This solution does not check.}$$

The solution of the equation $\sqrt{2x + 3} = x - 1$ is about 4.45.

For Exercises 3–10, solve. You may need to use the quadratic formula. Check to eliminate extraneous solutions.

3. $\sqrt{2y - 5} = 3$ **4.** $\sqrt[4]{4 - x} = 3$ **5.** $\sqrt{x + 3} = \sqrt{2x - 4}$

6. $3 + \sqrt[3]{x - 2} = 10$ **7.** $\sqrt{x - 3} = x - 3$ **8.** $\sqrt{4x + 1} = 2x - 1$

9. $\sqrt{2x + 13} = x + 5$ **10.** $\sqrt{x + 5} + 2 = 3x$

11. An equation for finding the wind chill temperature W is

$$W = 33 - (0.0393)(33 - T)(12.36 + 6.13\sqrt{v} - 0.32v)$$

where T is the temperature in °C and v is the wind velocity in kilometers per hour. Find each value for W in the table at the right. Round the answers to the nearest integer.

T	v	W
4°	15	?
4°	40	?
−2°	15	?
−2°	40	?

12. Open-ended In Example 1, the fraction $\dfrac{d}{1116}$ was used because the speed of sound in air is about 1116 ft/s. Do altitude and air temperature affect the speed of sound? Does sound travel at the same speed through air, water, and solids? Find information on the speed of sound and report your findings.

· · · · · · · · · · · · · · · · · · ·
Spiral Review

Simplify each radical expression. *(Section 8.2)*

13. $\sqrt[3]{-1728}$ **14.** $\sqrt{72}$ **15.** $\sqrt[5]{256}$

Tell whether each solution has *two solutions, one solution,* or *no solution*. *(Section 5.5)*

16. $y = 2x^2 + 10x - 28$ **17.** $y = 9x^2 + 24x + 16$

18. $y - 64 = x^2 - 16x$ **19.** $y = x^2 + 3x + 5$

Solve each system of equations. *(Section 7.1)*

20. $y = 6x - 5$
    $y = -9x$

21. $y = x^2 - 4x - 7$
    $y = -3x - 1$

Working with Complex Numbers

Learn how to . . .

- add, subtract, multiply, and divide complex numbers

- find complex solutions to equations that have no real solutions

So you can . . .

- solve problems involving complex numbers

Application

For electrical circuits with resistors, inductors, and capacitors, an important measurement for a circuit is called its *impedance*. You can use complex numbers to calculate the impedance in an electrical circuit.

Terms to Know   **Example / Illustration**

Terms to Know	Example / Illustration
Real number (p. 358) any number that is either rational or irrational	$32 \qquad 1.\overline{01} \qquad \dfrac{6}{7}$ $\sqrt{13} \qquad \pi$
Imaginary number (p. 359) any number of the form $a + bi$, where a and b are real numbers and $b \neq 0$ (The letter i represents the square root of -1, the fundamental unit in the system of imaginary numbers.)	$6 + 5i \qquad -3.2 - 5.7i$ $1 + \sqrt{17}i \qquad -\dfrac{1}{4}i$
Pure imaginary number (p. 359) the square root of any negative number	$\sqrt{-100} = 10i$ $\sqrt{-\dfrac{1}{4}} = \dfrac{1}{2}i$ $\sqrt{-2} = i\sqrt{2}$
Complex numbers (p. 359) the set formed by the imaginary numbers together with the real numbers	$3 + 7i \qquad -4i$ $15 - \sqrt{10}i \qquad 1.2$

Terms to Know	Example / Illustration
Complex conjugates (p. 361) two complex numbers of the form $a + bi$ and $a - bi$ (When complex conjugates are multiplied, the product is a real number.)	The complex conjugate of $5 + 3i$ is $5 - 3i$. The complex conjugate of $7 - 5i$ is $7 + 5i$.

UNDERSTANDING THE MAIN IDEAS

The square root of a negative number is called a *pure imaginary number*. Every pure imaginary number can be expressed in the form bi.

$$i = \sqrt{-1} \qquad i^2 = -1$$

Example 1

Express in bi form.

a. $\sqrt{-100}$ **b.** $\sqrt{-24}$ **c.** $\sqrt{-15}$

■ Solution ■

a. $\sqrt{-100} = \sqrt{-1}\,\sqrt{100} = (\sqrt{-1})(10) = 10i \quad \leftarrow i = \sqrt{-1}$

b. $\sqrt{-24} = \sqrt{-1}\,\sqrt{4}\,\sqrt{6} = 2i\sqrt{6}$

c. $\sqrt{-15} = \sqrt{-1}\,\sqrt{15} = i\sqrt{15}$

Express each pure imaginary number in bi form.

1. $\sqrt{-121}$ **2.** $\sqrt{-0.25}$ **3.** $\sqrt{-\dfrac{9}{49}}$ **4.** $\sqrt{-8}$

Operations with complex numbers

In a complex number $a + bi$, the value a is the real part and the value bi is the imaginary part.

The result of adding, subtracting, multiplying, or dividing two complex numbers is another complex number.

You add or subtract complex numbers by adding or subtracting their real parts and their imaginary parts separately.

$$(a + bi) + (c + di) = (a + c) + (b + d)i$$
$$(a + bi) - (c + di) = (a - c) + (b - d)i$$

You multiply two complex numbers just as you multiply two binomial expressions. *Complex conjugates* are used to divide complex numbers.

Example 2

Perform each operation. Express the result in $a + bi$ form.

a. $(3 - 17i) + (-12 - 22i)$ **b.** $(5 + 4i) - (15 - 20i)$ **c.** $(6 + 5i)(3 - 10i)$

d. $(4 + 3i)(4 - 3i)$ **e.** $\dfrac{8 - 5i}{2 + i}$ **f.** $(3 - 5i)^2$

■ Solution ■

a.
$$
\begin{aligned}
(3 - 17i) + (-12 - 22i) &= (3 + (-12)) + (-17i + (-22i)) \\
&= (3 - 12) + (-17i - 22i) \\
&= -9 - 39i
\end{aligned}
$$

b.
$$
\begin{aligned}
(5 + 4i) - (15 - 20i) &= (5 - 15) + (4i - (-20i)) \\
&= (5 - 15) + (4i + 20i) \\
&= -10 + 24i
\end{aligned}
$$

c.
$$
\begin{aligned}
(6 + 5i)(3 - 10i) &= 6(3 - 10i) + 5i(3 - 10i) \\
&= (6)(3) + (6)(-10i) + (5i)(3) + (5i)(-10i) \\
&= 18 - 60i + 15i - 50i^2 \\
&= 18 - 45i - 50(-1) \qquad \leftarrow\ i^2 = -1 \\
&= 68 - 45i
\end{aligned}
$$

d.
$$
\begin{aligned}
(4 + 3i)(4 - 3i) &= (4)(4) + (4)(-3i) + (3i)(4) + (3i)(-3i) \\
&= 16 - 12i + 12i - 9i^2 \\
&= 16 - 9(-1) \\
&= 25, \text{ or } 25 + 0i
\end{aligned}
$$

e. In part (d) above, notice that the product of two complex conjugates is a real number. When two complex numbers are divided, this fact is used to change the complex denominator into a real number. We begin here by multiplying both the numerator and denominator by the complex conjugate of the denominator, $2 - i$.

$$
\begin{aligned}
\frac{8 - 5i}{2 + i} &= \frac{8 - 5i}{2 + i} \cdot \frac{2 - i}{2 - i} \\[2mm]
&= \frac{(8)(2) + (8)(-i) + (-5i)(2) + (-5i)(-i)}{(2)(2) + (2)(-i) + (i)(2) + (i)(-i)} \\[2mm]
&= \frac{16 - 8i - 10i + 5i^2}{4 - 2i + 2i - i^2} \\[2mm]
&= \frac{16 - 18i + 5(-1)}{4 - (-1)} \\[2mm]
&= \frac{11 - 18i}{5} \\[2mm]
&= \frac{11}{5} - \frac{18}{5i}
\end{aligned}
$$

f.
$$
\begin{aligned}
(3 - 5i)^2 &= (3 - 5i)(3 - 5i) \\
&= (3)(3) + (3)(-5i) + (-5i)(3) + (-5i)(-5i) \\
&= 9 - 15i - 15i + 25i^2 \\
&= 9 - 30i + 25(-1) \\
&= -16 - 30i
\end{aligned}
$$

Add, subtract, or multiply.

5. $(12 - 8i) + (-10 - i)$ **6.** $(-5 + i) - (7 - 3i)$ **7.** $(-4i)(-8i)$

8. $3i(5 - 2i)$ **9.** $6(5i - 2)$ **10.** $(2 + 3i)(2 - 3i)$

11. $\left(\dfrac{1}{2} + \dfrac{1}{4}i\right)\left(\dfrac{1}{2} - \dfrac{1}{4}i\right)$ **12.** $(5 - i)^2$ **13.** $(2 + 5i)(3 - 8i)$

Use complex conjugates to express each quotient in $a + bi$ form.

14. $\dfrac{4 + 2i}{3 - i}$ **15.** $\dfrac{5 - 4i}{5 + 4i}$ **16.** $\dfrac{7 + 4i}{-3 + 2i}$

Solving quadratic equations with complex roots

You can use the quadratic formula, $x = \dfrac{-b \pm \sqrt{b^2 - 4ac}}{2a}$, to find the complex roots of a quadratic equation.

Example 3

Solve each equation. Check the solutions.

   a. $x^2 - 4x + 5 = 0$ **b.** $x^2 + 20x + 116 = 0$

■ Solution ■

   a. Use the quadratic formula with $a = 1$, $b = -4$, and $c = 5$.

$$x = \frac{-(-4) \pm \sqrt{(-4)^2 - 4(1)(5)}}{2(1)}$$

$$= \frac{4 \pm \sqrt{16 - 20}}{2}$$

$$= \frac{4 \pm \sqrt{-4}}{2}$$

$$= \frac{4 \pm 2i}{2}$$

$$= 2 \pm i$$

Check both solutions:

$(2 + i)^2 - 4(2 + i) + 5 \overset{?}{=} 0$ $(2 - i)^2 - 4(2 - i) + 5 \overset{?}{=} 0$

$4 + 4i + i^2 - 8 - 4i + 5 \overset{?}{=} 0$ $4 - 4i + i^2 - 8 + 4i + 5 \overset{?}{=} 0$

            $1 + i^2 \overset{?}{=} 0$ $1 + i^2 \overset{?}{=} 0$

          $1 + (-1) \overset{?}{=} 0$ $1 + (-1) \overset{?}{=} 0$

                 $0 = 0$ ✓ $0 = 0$ ✓

(Solution continues on next page.)

■ **Solution** ■ *(continued)*

b. Use the quadratic formula with $a = 1$, $b = 20$, and $c = 116$.

$$x = \frac{-(20) \pm \sqrt{(20)^2 - 4(1)(116)}}{2(1)}$$

$$= \frac{-20 \pm \sqrt{400 - 464}}{2}$$

$$= \frac{-20 \pm \sqrt{-64}}{2}$$

$$= \frac{-20 \pm 8i}{2}$$

$$= -10 \pm 4i$$

Check both solutions:

$$(-10 + 4i)^2 + 20(-10 + 4i) + 116 \stackrel{?}{=} 0$$
$$100 - 80i + 16i^2 - 200 + 80i + 116 \stackrel{?}{=} 0$$
$$16 + 16i^2 \stackrel{?}{=} 0$$
$$16 + 16(-1) \stackrel{?}{=} 0$$
$$0 = 0 \quad \checkmark$$

$$(-10 - 4i)^2 + 20(-10 - 4i) + 116 \stackrel{?}{=} 0$$
$$100 + 80i + 16i^2 - 200 - 80i + 116 \stackrel{?}{=} 0$$
$$16 + 16i^2 \stackrel{?}{=} 0$$
$$16 + 16(-1) \stackrel{?}{=} 0$$
$$0 = 0 \quad \checkmark$$

For Exercises 17–20, solve using the quadratic formula. Check your solutions.

17. $x^2 - 6x + 13 = 0$

18. $x^2 - 10x + 26 = 0$

19. $x^2 + 14x = -58$

20. $x^2 + 9x = -9x - 82$

21. For two electric pathways wired as parallel circuits, the total impedance Z is the product of the impedance for each pathway ($Z_1 Z_2$) divided by their sum ($Z_1 + Z_2$). Use the formula $Z = \dfrac{Z_1 Z_2}{Z_1 + Z_2}$ to find the total impedance if the impedances of the two pathways are $5 + 3i$ ohms and $2 - 4i$ ohms, respectively.

Solve. *(Section 8.1)*

22. $\sqrt{x + 5} = 3$ **23.** $\sqrt[4]{3x - 2} = 3$ **24.** $\sqrt[3]{3x + 4} = 4$

Write an exponential function whose graph passes through each pair of points.
(Section 3.5)

25. $(2, 8), (5, 125)$ **26.** $(4, 8), (6, 18)$ **27.** $(1, 1), (2, 32)$

Use the Pythagorean theorem to find the length of the hypotenuse of each right triangle. *(Toolbox, p. 801)*

28.

29.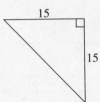

30.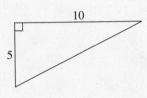

GOAL

Visualizing the Complex Plane

Learn how to ...

- plot complex numbers in the complex plane

- calculate the magnitude of a complex number

So you can ...

- understand the mathematics used to generate images of points in the complex plane

Application

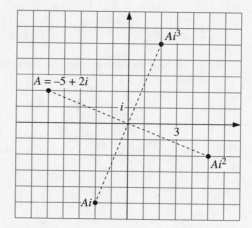

Computer animations are built using rotations and other movements of points on a grid. You can use a point in the complex plane to represent a complex number; for example, point A above models $-5 + 2i$. Then you can rotate that point about the origin, in quarter-turn increments, by multiplying by the powers of i.

Terms to Know

Example / Illustration

Terms to Know	Example / Illustration
Complex plane (p. 635) a coordinate system with real numbers on the horizontal axis and imaginary numbers on the vertical axis, where the point (a, b) represents the complex number $a + bi$	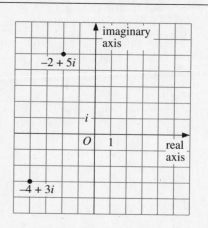
Imaginary axis (p. 365) the vertical axis of a complex plane	
Real axis (p. 365) the horizontal axis of a complex plane	
Magnitude (p. 366) for a complex number $a + bi$, the distance from $(0, 0)$ to (a, b) (The magnitude of the complex number is denoted by $\lvert a + bi \rvert$; $\lvert a + bi \rvert = \sqrt{a^2 + b^2}$.)	$\lvert -2 + 5i \rvert = \sqrt{(-2)^2 + 5^2}$ $= \sqrt{4 + 25}$ $= \sqrt{29}$

UNDERSTANDING THE MAIN IDEAS

In the complex plane, the point (a, b) represents the complex number $a + bi$. The coordinate a represents the distance along the horizontal, real axis and the coordinate b represents the distance along the vertical, imaginary axis.

The magnitude of the complex number $a + bi$, denoted by $|a + bi|$, is $\sqrt{a^2 + b^2}$. The magnitude of a complex number is a distance, so it is always a positive real number.

Example 1

Use the complex numbers $5 - 5i$, $-4 - i$, $6i$, and $8 + 2i$.

a. Plot the complex numbers on the same complex plane.

b. Find the magnitude of each complex number.

▪ Solution ▪

a. The ordered pairs for the complex numbers are $(5, -5)$, $(-4, -1)$, $(0, 6)$, and $(8, 2)$, respectively. The points are graphed in the figure below.

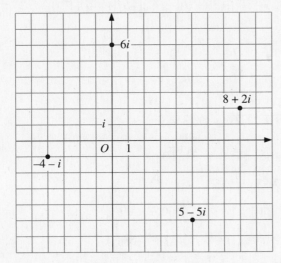

b. $|5 - 5i| = \sqrt{5^2 + (-5)^2} = \sqrt{50} = \sqrt{25 \cdot 2} = 5\sqrt{2}$

$|-4 - i| = \sqrt{(-4)^2 + (-1)^2} = \sqrt{17}$

$|6i| = |0 + 6i| = \sqrt{0^2 + 6^2} = \sqrt{36} = 6$

$|8 + 2i| = \sqrt{8^2 + 2^2} = \sqrt{68} = \sqrt{4 \cdot 17} = 2\sqrt{17}$

Plot each complex number. Then find its magnitude.

1. $6 - 8i$

2. $-13i$

3. $4 + 4\sqrt{3}i$

4. $-4 + 5i$

5. $-5 - 12i$

6. $-15 + 8i$

You can discover an interesting property of complex numbers when you calculate the magnitude of the product of two complex numbers and the product of their magnitudes.

Example 2

For each pair of complex numbers, calculate their magnitudes, the product of their magnitudes, their product, and the magnitude of their product. Do you see a pattern in the results?

   a. $3 + 4i, -5 + 12i$ **b.** $-2 - 6i, 5 + i$ **c.** $4i, 2 - 3i$

■ Solution ■

a. The magnitudes are:
$$|3 + 4i| = \sqrt{3^2 + 4^2} = \sqrt{25} = 5$$
$$|-5 + 12i| = \sqrt{(-5)^2 + 12^2} = \sqrt{169} = 13$$

The product of the magnitudes is $(5)(13) = 65$.

Their product is:
$$(3 + 4i)(-5 + 12i) = (3)(-5) + (3)(12i) + (4i)(-5) + (4i)(12i)$$
$$= -15 + 36i - 20i + 48i^2$$
$$= -15 + 16i + 48(-1)$$
$$= -63 + 16i$$

The magnitude of their product is:
$$|-63 + 16i| = \sqrt{(-63)^2 + 16^2} = \sqrt{4225} = 65$$

b. $|-2 - 6i| = \sqrt{(-2)^2 + (-6)^2} = \sqrt{40} = 2\sqrt{10}$

   $|5 + i| = \sqrt{5^2 + 1^2} = \sqrt{26}$

product of magnitudes: $2\sqrt{10}\sqrt{26} = 2\sqrt{260} = 4\sqrt{65}$

$$(-2 - 6i)(5 + i) = (-2)(5) + (-2)(i) + (-6i)(5) + (-6i)(i)$$
$$= -10 - 2i - 30i - 6i^2$$
$$= -10 - 32i - 6(-1)$$
$$= -4 + 32i$$

magnitude of product: $|-4 + 32i| = \sqrt{(-4)^2 + 32^2}$
$$= \sqrt{1040}$$
$$= 4\sqrt{65}$$

(Solution continues on next page.)

■ **Solution** ■ *(continued)*

c. $|4i| = \sqrt{0^2 + 4^2} = \sqrt{16} = 4$

$|2 - 3i| = \sqrt{2^2 + (-3)^2} = \sqrt{13}$

product of magnitudes: $4\sqrt{13}$

$$(4i)(2 - 3i) = (4i)(2) + (4i)(-3i)$$
$$= 8i - 12i^2$$
$$= 8i - 12(-1)$$
$$= 12 + 8i$$

magnitude of product: $|12 + 8i| = \sqrt{12^2 + 8^2}$
$$= \sqrt{144 + 64}$$
$$= \sqrt{208}$$
$$= 4\sqrt{13}$$

For all three pairs of complex numbers in parts (a)–(c), the product of the magnitudes is equal to the magnitude of the product.

For each pair of complex numbers in Exercises 7–12, find:
  a. their magnitudes.
  b. the product of their magnitudes.
  c. their product.
  d. the magnitude of their product.

7. $6 - 10i, -4 + 3i$ **8.** $2 + 2i, -4 + i$ **9.** $i, -5i$

10. $10 - i, 3i$ **11.** $5 - 3i, -5 + 3i$ **12.** $2 + 6i, 2 - 6i$

13. A complex number and its *opposite* have a sum of 0. What is the opposite of $-7 + 12i$? the opposite of $x + yi$?

14. Recall that the *complex conjugate* of $a + bi$ is $a - bi$. Plot the number $2 + 3i$ and its complex conjugate in the complex plane. Then plot their sum.

15. Open-ended Choose two complex numbers and find their magnitude, their sum, and the sum of their magnitudes. Do you think there is a pattern similar to the one shown in Example 2? If so, explain the pattern.

16. Mathematics Journal Write answers to these questions: Can the magnitude of a complex number ever be a negative integer? When is the magnitude of a complex number an integer? When is the magnitude of a complex number an irrational number?

Refer to the Application at the beginning of this section.

17. Find the products $i(-5 + 2i)$, $i^2(-5 + 2i)$, and $i^3(-5 + 2i)$.

18. Find the magnitude of $-5 + 2i$, as well as the magnitude of each product found in Exercise 17. How do these magnitudes compare?

Find the mean and standard deviation of each data set. *(Section 6.5)*

	Fill-ins (%)	Multiple-choice (%)	Essay (%)
State Math exam	52	40	8
State History exam	30	60	10
State Biology exam	35	60	5
State Chemistry exam	47	50	3
State Language exam	45	40	15

19. percentage of fill-in questions

20. percentage of multiple-choice questions

21. percentage of essay questions

Write a quadratic equation with the given solutions. *(Section 8.4)*

22. $2i, -2i$

23. $i\sqrt{5}, -i\sqrt{5}$

Evaluate each matrix expression. *(Sections 1.3 and 1.4)*

24. $\begin{bmatrix} 2 & 5 \\ 0 & -3 \end{bmatrix} + \begin{bmatrix} 8 & -5 \\ 6 & 4 \end{bmatrix}$

25. $\begin{bmatrix} 2 & 5 \\ 0 & -3 \end{bmatrix}\begin{bmatrix} 8 & -5 \\ 6 & 4 \end{bmatrix}$

26. $\begin{bmatrix} 7 & 5 \\ 4 & 3 \end{bmatrix}\begin{bmatrix} 3 & -5 \\ -4 & 7 \end{bmatrix}$

27. $\begin{bmatrix} 3 & 5 & 6 \\ 1 & -2 & 1 \\ 4 & 3 & -6 \end{bmatrix} + \begin{bmatrix} 2 & -5 & 1 \\ -2 & -2 & -2 \\ 1 & 0 & -1 \end{bmatrix}$

GOAL

Properties of Number Systems

Learn how to . . .

- identify the number systems to which a number belongs
- evaluate whether group properties hold for a set and an operation

So you can . . .

- see structural similarities in groups

Application

The road mileage between two cities is the same in each direction. You can use this property to fill in the missing numbers in the mileage table shown below.

	Atl.	Chi.	Denv.	Mem.	Minn.
Atlanta		708	?	?	1121
Chicago	?		1021	?	410
Denver	1430	?		1043	920
Memphis	382	537	?		?
Minneapolis	?	?	?	914	

Terms to Know	*Example / Illustration*
Group (p. 373) a number set and operation together, if the set is closed under the operation and the *identity, inverse* and *associativity properties* hold	The whole numbers and the operation of multiplication form a group.
Commutative group (p. 373) a number set and operation that form a group and for which the *commutative property* also holds	The even integers and the operation of addition form a commutative group.
Closure property (p. 373, 374) the property that states that the result of an operation on any two members of a set is also in that set	Since $5 + 8 = 13$ and 13 is a positive integer, the set of positive integers is closed under addition. However, since $5 - 8 = -3$ and -3 is not a positive integer, the set of positive integers is *not* closed under subtraction.

Terms to Know	**Example / Illustration**
Identity property (p. 373, 374) the property that states there is a member *I* in a set such that an operation by *I* on any other member of the set results in the member you started with	The number 0 is the identity for the real numbers under addition. The number 1 is the identity for the real numbers under multiplication.
Inverse property (p. 373, 374) the property that states that for every member in a set there is another member such that the result of a specific operation on those two members results in the identity for that set and operation	Under addition, any real number *n* has an inverse $(-n)$ because $n + (-n) = 0$, the identity under addition for the set of real numbers. $$5 + (-5) = 0;\ \frac{3}{4} + \left(-\frac{3}{4}\right) = 0$$ Under multiplication, any nonzero real number *m* has an inverse $\frac{1}{m}$ because $m \cdot \frac{1}{m} = 1$, the identity under multiplication for the set of nonzero real numbers. $$(5)\left(\frac{1}{5}\right) = 1;\ \left(-\frac{3}{4}\right)\left(-\frac{4}{3}\right) = 1$$
Associative property (p. 373, 374) the property that states that if you perform an operation on two members and then perform that same operation on the result and a third member, this final result is the same as if you perform the operation on the second and third members and then again on the first member and the result	For the set of real numbers, addition and multiplication are associative. $$(5 + 3) + 4 = 5 + (3 + 4)$$ $$(5 \cdot 3) \cdot 4 = 5 \cdot (3 \cdot 4)$$ For the set of real numbers, subtraction and division are *not* associative. $$(5 - 3) - 4 \neq 5 - (3 - 4)$$ $$(5 \div 3) \div 4 \neq 5 \div (3 \div 4)$$
Commutative property (p. 373, 374) the property that states the order in which you perform an operation on two members of a set does not affect the result	For the set of real numbers, addition and multiplication are commutative. $$5 + (-7) = (-7) + 5$$ $$5 \cdot (-7) = (-7) \cdot 5$$ For the set of real numbers, subtraction and division are *not* commutative.

206

UNDERSTANDING THE MAIN IDEAS

Nested sets of numbers

The complex number system is made of many different kinds of numbers. The diagram below is one way to show how the various important number systems are "nested." A number located in any box belongs to the number systems represented by all the boxes surrounding it.

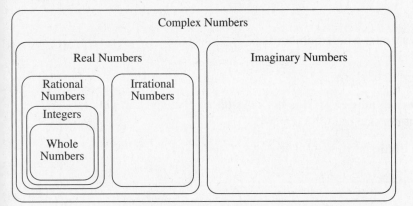

Example 1

Identify the number systems to which each number belongs.

a. $\sqrt{5}$ **b.** $-\sqrt{9}$ **c.** $5 - 3i$ **d.** 1 **e.** $5i$ **f.** i^2

■ Solution ■

a. The number $\sqrt{5}$ belongs to the irrational, real, and complex number systems.

b. The number $-\sqrt{9} = -3$, so it belongs to the integer, rational, real, and complex number systems.

c. The number $5 - 3i$ belongs to the complex number system.

d. The number 1 belongs to the whole, integer, rational, real, and complex number systems.

e. The number $5i$ belongs to the imaginary and complex number systems.

f. The number $i^2 = -1$ so it belongs to the integer, rational, real, and complex number systems.

Example 2

If possible, give an example of a number that satisfies each description.

a. integral and whole   **b.** real and rational

c. whole and but not integral   **d.** real and irrational

■ Solution ■

a. A number that is an integer and a whole number is 5.

b. A number that is real and rational is $\frac{2}{3}$.

c. Every whole number is also an integer, so it is *not possible* to give an example of a number that is whole but not integral.

d. A number that is real and irrational is $\sqrt{3}$.

Identify the number systems to which each number belongs.

1. $-\sqrt{5}$ **2.** $3i$ **3.** $\frac{10}{3}$

4. $\sqrt{-9}$ **5.** $4 - 2i$ **6.** 11

If possible, give an example of a number that satisfies each description.

7. rational and whole   **8.** rational but not integral

9. real and imaginary   **10.** integral but not whole

Properties of groups

Different number systems have different properties. Two important questions for any number system are "Is it closed under a particular operation?" and "If it is closed, does it form a group?".

Example 3

Tell whether each number set is closed under the given operation. If so, tell whether the number set and operation *form a group*, *form a commutative group*, or *do not form a group*. Explain.

a. negative rational numbers; $\times$ **b.** even integers; $+$

c. integers; $\times$ **d.** numbers $-10, 0, 10$; $+$

■ Solution ■

a. The product of any two negative rational numbers is a positve rational number, so the set is *not* closed under multiplication. So the set of negative rational numbers and multiplication *do not form a group*.

b. The result of adding any two even integers is another even integer, so this set is closed under addition. The set of even integers contains 0, the identity element for addition. The set contains an additive inverse $-a$ for every element a, so that $-a + a = a + (-a) = 0$. The set is both associative and commutative under addition. So the set of even integers and addition *form a commutative group*.

c. The product of any two integers is always an integer, so the integers are closed under multiplication. There is an identity element, 1, for the set under multiplication. However, the only integers that have inverse elements in the set are -1 and 1, so the set of integers and multiplication *do not form a group*.

d. For the set of numbers -10, 0, 10, if you add 10 to itself you do not get an element of the set. Therefore, the set is not closed under addition. So the set of numbers -10, 0, and 10 and the operation addition *do not form a group*.

Tell whether each number set is closed under the given operation. If so, tell whether the number set and operation *form a group, form a commutative group*, or *do not form a group*. Explain.

11. positive rational numbers; $\times$

12. irrational numbers; $+$

13. real numbers; $+$

14. negative real numbers; $\times$

15. complex numbers; $+$

16. -1, 0, 1; $\times$

A *field* consists of a set of elements and two operations (usually addition and multiplication) that satisfy all the commutative group properties, as well as the following property:

$$\text{for all } a, b, \text{ and } c \text{ in the group, } a \times (b + c) = (a \times b) + (a \times c).$$

Use this information for Exercises 17 and 18.

17. What is the name usually given to this property?

18. Do the complex numbers form a field? Do the irrational numbers form a field?

19. Copy the chart from the Application at the beginning of this section. Fill in the missing numbers in the chart.

20. Mathematics Journal Which cities from the chart in Exercise 19 have you seen in person or in movies? Describe your impressions of each city.

Plot each pair of complex numbers and their product in the complex plane.
(Section 8.5)

21. $5 + i$, $-4 - 3i$

22. $3i$, $2 + 5i$

Write each quadratic function as a product of factors. *(Section 5.6)*

23. $y = 2x^2 + x - 3$

24. $y = 25x^2 + 20x + 3$

Simplify. *(Toolbox, p. 781)*

25. $(3x - 7) + (2x - 5)$

26. $(3x + 5) - (4 - 9x)$

Chapter 8 Review ··············

Complete these exercises for a review of Chapter 8. If you have difficulty with a particular problem, review the indicated section.

Graph each function. State the domain and range. *(Section 8.1)*

1. $y = \frac{1}{2}x^2$

2. $y = \sqrt{x} - 3$

3. the inverse of
$$f(x) = \frac{x^2}{4}, x \le 0$$

Simplify each radical expression. *(Section 8.2)*

4. $\sqrt[3]{-54}$

5. $\sqrt[4]{\frac{1}{81}}$

6. $\sqrt[3]{0.216}$

Express using fractional exponents. *(Section 8.2)*

7. $\sqrt[5]{a^{10}}$

8. $\sqrt[3]{b^2}$

9. $\sqrt[4]{x^3 y}$

Solve each radical equation. *(Section 8.3)*

10. $\sqrt{2x + 5} = 1$

11. $3 + \sqrt{3x + 5} = 12$

12. $\sqrt{x - 2} = x - 4$

For Exercises 13–18, perform each operation. Express the result in *a + bi* form. *(Section 8.4)*

13. $(3 - 7i) + (-5 + 3i)$

14. $(-5 + 8i) - (-6 - 3i)$

15. $(2 + 5i)(2 - 5i)$

16. $(3 - 2i)(4 - i)$

17. $\frac{2 + 7i}{5 - 2i}$

18. $\frac{13 + 3i}{5i}$

19. Solve $x^2 - 6x + 11 = 0$ using the quadratic formula. *(Section 8.4)*

For Exercises 20–22, use the complex numbers –2 – 2*i* and 4 + *i*. *(Section 8.5)*

20. Find the magnitude of each complex number.

21. Find the product of the two complex numbers.

22. What is the magnitude of the product you found in Exercise 21?

23. Writing Describe the similarities and differences between graphing an ordered pair in a coordinate plane and graphing a complex number in the complex plane. *(Section 8.5)*

Identify the number systems to which each number belongs. *(Section 8.6)*

24. $-3i$

25. $\frac{14}{3}$

26. π

27. $2 - 5\sqrt{3}i$

28. 1

29. $\sqrt{21}$

Tell whether each number set is closed under the given operation. If so, tell whether the number set and operation *form a group, form a commutative group,* or *do not form a group.* Explain. *(Section 8.6)*

30. odd integers; $\times$

31. odd integers; $+$

32. complex numbers; $\times$

33. rational numbers; $+$

SPIRAL REVIEW   Chapters 1–8

For Exercises 1–3, use $R = \begin{bmatrix} 4 & 2 \\ 3 & 1 \end{bmatrix}$, $S = \begin{bmatrix} 1 & 0 \\ 0 & 1 \end{bmatrix}$, and $T = \begin{bmatrix} 5 & -3 \\ 1 & 0 \end{bmatrix}$.

1. Find $RS + SR$.

2. Find $-3S$.

3. Is $RT = TR$?

Find the equation, in slope-intercept form, for each line.

4. through $(-3, 7)$ and $(4, -7)$

5. through $(6, 6)$ and $(-10, -2)$

Evaluate each expression.

6. $8^{5/3}$

7. $\log_{64} 8$

8. $\log_{16} 64$

Solve for *x*.

9. $\log_2 32 = x$

10. $\log_x 32 = \dfrac{5}{4}$

11. $x^2 + 3 = 39$

12. $\dfrac{1}{x^2} = \dfrac{1}{100}$

13. $x^2 - 10x + 21 = 0$

14. $x^2 + x - 8 = 0$

For Exercises 15 and 16, use the function $f(x) = 3x^2 + 12x + 11$.

15. Write the function in vertex form, $y = a(x - h)^2 + k$.

16. What is the minimum value of the function?

17. Which set of data has a greater mean value: 12, 13, 15, 16, 20 or 14.8, 14.9, 15.3, 15.4, 15.8?

Graph each system.

18. $y = -2x + 7$
   $y = \dfrac{1}{5}x - 4$

19. $y < x + 3$
   $y < -x + 5$

Polynomials

Learn how to ...

- recognize, evaluate, add, and subtract polynomials

So you can ...

- understand numeration systems

Application

Each year on her birthday, Lisa's grandparents deposit $200 into a savings account that pays interest annually. After making the fifth yearly deposit, the total amount in the account is given by the polynomial expression

$$200x^4 + 200x^3 + 200x^2 + 200x + 200$$

where x represents 1 plus the annual interest rate.

Terms to Know	**Example / Illustration**
Polynomial (p. 390) a term or sum or terms, where each term is the product of a real-number coefficient and a variable with a whole-number exponent	$3x^5 \qquad -17x^5 + 3x^4 - 1$ (*Note*: The expression $4x^3 + 5ix^2$ is *not* a polynomial because the coefficient of x^2 is not a real number; the expression $-17x^4 + \sqrt{13}x^3 - 2x^{1/2}$ is *not* a polynomial because the exponent for the term $2x^{1/2}$ is not a whole number.)
Degree (p. 390) the greatest exponent of a polynomial	The degree of each polynomial below is 5. $3x^5 + 100x^4 - 17x^3 + x^2 - 2x$ $15x^5 - 2$ $10x^2 + 15x^4 - x^5$
Constant term (p. 392) the term of a polynomial that appears without a variable	The constant term in the polynomial $17x^3 + x^2 - 2x + 7$ is 7. (*Note*: The constant term 7 is the same as $7x^0$, since $x^0 = 1$.)
Standard form (p. 390) a polynomial written so that its exponents decrease from left to right	The standard form of the polynomial $10x^2 + 15x^4 - x^5$ is $-x^5 + 15x^4 + 10x^2$.

Study Guide, ALGEBRA 2: EXPLORATIONS AND APPLICATIONS

213

Terms to Know	Example / Illustration
Synthetic substitution (p. 391) a shortcut procedure for evaluating a polynomial for a particular value of the variable	To evaluate $2x^3 - 3x^2 + x + 5$ for $x = 4$: 1. Write the value of x followed by the coefficients of the polynomial on one line, as shown: <u>4</u> 2 −3 1 5 2. Bring down the first coefficient. Now multiply the number brought down by the value of x and add the product to the next coefficient. Bring the sum down and repeat this step until you obtain the last sum. 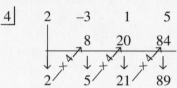 3. The value of the polynomial for the given value of x is the last sum found. The value of $2x^3 - 3x^2 + x + 5$ for $x = 4$ is 89.
Like terms (p. 392) terms that have the same power of the same variable	The terms $13x^4$, $-7x^4$, and x^4 are like terms. (The terms $7x^3$ and $7x^5$ are *not* like terms; neither are $6x^2$ and $7y^2$.)

UNDERSTANDING THE MAIN IDEAS

Describing polynomials

It is important to recognize when an expression is a polynomial.

Example 1

Tell whether each expression is a polynomial. If so, write the polynomial in standard form and state its degree. If not, explain why not.

a. $3r^2 + 4r^3 - \frac{1}{2}r$

b. $5y^3 - 2y^2 + y - 8 + 3y^{-1}$

c. $5.3s^4 + \sqrt{5}s - 6 + 4.7s^2$

d. $\frac{1}{3}x^2 + \frac{1}{5}x + \frac{3}{x-2}$

e. $m^5 + 11$

f. $n^2 + \sqrt{-4}n$

■ **Solution** ■

a. The expression *is* a polynomial. Its standard form is $4r^3 + 3r^2 - \frac{1}{2}r$, and its degree is 3.

b. The expression *is not* a polynomial, because the exponent of the term $3y^{-1}$ is not a whole number.

c. The expression *is* a polynomial. Its standard form is $5.3s^4 + 4.7s^2 + \sqrt{5}s - 6$, and its degree is 4.

d. The expression *is not* a polynomial, because $\frac{3}{x-2}$ is not the product of a real number and a whole-number power of x.

e. The expression *is* a polynomial. It is in standard form, and its degree is 5.

f. The expression *is not* a polynomial, because the coefficient of the term $\sqrt{-4}n$ is not a real number.

Tell whether each expression is a polynomial. If so, write the polynomial in standard form and state its degree. If not, explain why not.

1. $10x^2 + 9x^3 + 8x^4 - 3$

2. $\frac{1}{2}y^3 - \frac{2}{3}y^4 - \frac{3}{4}y + \frac{1}{8}$

3. $5x^{-3} - 3x^{-2} + 4x^{-1}$

4. $\sqrt{2x} + \sqrt{2}x$

5. $\sqrt{5}x^3 + \sqrt{3}x^2 + \sqrt{2}x$

6. $8 - 15m^{15}$

Adding and subtracting polynomials

To add two polynomials, you add the coefficients of *like* terms of the polynomials. To subtract a polynomial, you add the opposite of each term of that polynomial to its like term in the first polynomial. (*Note*: It is helpful to rewrite the polynomials in standard form if they are not given it this form.)

Example 2

Find each sum or difference.

a. $(7x^3 - 8x^2 + 12x - 3) + (10x^3 - 5x + 1)$

b. $(10x^4 + 3x^3 - 4x^2 - 5x + 1) - (3x^4 + 5x^3 - 2x^2 + 6x + 1)$

■ Solution ■

a. Align like terms vertically before adding their coefficients. Notice that there is no x^2-term in the second polynomial.

$$
\begin{array}{rrrrr}
7x^3 & - 8x^2 & + 12x & - 3 & \\
+\ 10x^3 & + 0x^2 & - 5x & + 1 & \leftarrow \text{Use } 0x^2 \text{ to hold the place of the } x^2\text{-term.} \\
\hline
17x^3 & - 8x^2 & + 7x & - 2 &
\end{array}
$$

The sum is $17x^3 - 8x^2 + 7x - 2$.

b. Align like terms vertically.

$$
\begin{array}{rrrrr}
10x^4 & + 3x^3 & - 4x^2 & - 5x & + 1 \\
-\ (3x^4 & + 5x^3 & - 2x^2 & + 6x & + 1) \\
\end{array}
$$

Now write the opposite of each term of the subtracted polynomial and then add.

$$
\begin{array}{rrrrr}
10x^4 & + 3x^3 & - 4x^2 & - 5x & + 1 \\
+\ (-3x^4 & - 5x^3 & + 2x^2 & - 6x & - 1) \\
\hline
7x^4 & - 2x^3 & - 2x^2 & - 11x &
\end{array}
$$

The difference is $7x^4 - 2x^3 - 2x^2 - 11x$.

Add or subtract.

7. $(3x^4 + 5x^3 + 3x^2 - x + 1) + (-5x^4 + 4x^3 - 2x^2 + 3x - 5)$

8. $(3x^4 + 5x^3 - 3x^2 + x + 1) - (-5x^4 + 4x^3 - 2x^2 + 3x - 5)$

9. $(1 + y^2 + 7y^3) + (5 + 6y - 2y^2 - y^3)$

10. $(5 + 6y - 2y^2 - y^3) - (1 + y^2 + 7y^3)$

Synthetic substitution

You can evaluate a polynomial for a particular value of the variable by using a procedure called *synthetic substitution*.

Example 3

Use synthetic substitution to evaluate each polynomial for the given value of the variable.

a. $3x^4 - 5x^3 - 2x^2 + 6x + 1$; $x = 4$ **b.** $7x^5 + 8x^4 + 12x - 3$; $x = -2$

■ Solution ■

a. Write the given value of x, 4, and the coefficients of the polynomial in a row. Then bring down the first coefficient, 3, multiply it by 4, and add the product to the next coefficient, –5. Repeat this process of multiplying and adding for each of the remaining coefficients. The value of the polynomial expression for $x = 4$ is the last sum found in this procedure.

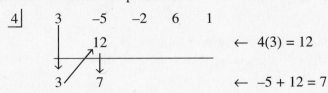

$$
\begin{array}{r|rrrrr}
4 & 3 & -5 & -2 & 6 & 1 \\
  &   & 12 &   &   &   \\
\hline
  & 3 & 7 &   &   &   \\
\end{array}
$$
$\leftarrow 4(3) = 12$
$\leftarrow -5 + 12 = 7$

$$
\begin{array}{r|rrrrr}
4 & 3 & -5 & -2 & 6 & 1 \\
  &   & 12 & 28 &   &   \\
\hline
  & 3 & 7 & 26 &   &   \\
\end{array}
$$
$\leftarrow 4(7) = 28$
$\leftarrow -2 + 28 = 26$

$$
\begin{array}{r|rrrrr}
4 & 3 & -5 & -2 & 6 & 1 \\
  &   & 12 & 28 & 104 & \\
\hline
  & 3 & 7 & 26 & 110 & \\
\end{array}
$$
$\leftarrow 4(26) = 104$
$\leftarrow 6 + 104 = 110$

$$
\begin{array}{r|rrrrr}
4 & 3 & -5 & -2 & 6 & 1 \\
  &   & 12 & 28 & 104 & 440 \\
\hline
  & 3 & 7 & 26 & 110 & 441 \\
\end{array}
$$
$\leftarrow 4(110) = 440$
$\leftarrow 1 + 440 = 441$

The value of $3x^4 - 5x^3 - 2x^2 + 6x + 1$ for $x = 4$ is 441.

b. Notice that this polynomial does not have an x^3-term or an x^2-term. So when writing the given value of x, –2, and the coefficients of the polynomial in a row, you must write a zero as the coefficient of each of these terms.

$$
\begin{array}{r|rrrrrr}
-2 & 7 & 8 & 0 & 0 & 12 & -3 \\
   &   & -14 & 12 & -24 & 48 & -120 \\
\hline
   & 7 & -6 & 12 & -24 & 60 & -123 \\
\end{array}
$$
$\leftarrow$ Notice the two zeros.

The value of $7x^5 + 8x^4 + 12x - 3$ when $x = -2$ is –123.

Use synthetic substitution to evaluate each polynomial for the given value of the variable.

11. $-y^3 - 2y^2 + 6y + 5$ for $y = -3$

12. $-5x^4 + 4x^3 - 2x^2 + 3x - 5$ for $x = 3$

13. $3x^4 + 5x^2 + 1$ for $x = 2$

14. $6y + 5 - y^3 - 2y^2$ for $y = 4$

TECHNOLOGY Refer to the Application at the beginning of this section. Suppose the annual interest rate paid by the bank is 6%, so $x = 1.06$.

15. Use a calculator to evaluate $200x^4$ for $x = 1.06$, which is the value of just the original \$200 deposit (without compounding) after four full years.

16. Use a calculator to find the total value of Lisa's account after the fifth deposit.

....

Spiral Review

Tell whether each set of numbers is closed under the given operation. If so, tell whether the number set and operation *form a group,* **form a commutative** *group,* **or** *do not form a group.* **Explain.** *(Section 8.6)*

17. imaginary numbers; $\times$

18. complex numbers; $+$

19. odd numbers; $+$

20. numbers of the form 2^n, where n is an integer; $\times$

Evaluate each logarithm. *(Section 4.2)*

21. $\log_5 125$

22. $\log 1{,}000{,}000$

23. $\log_{27} 9$

Multiply. *(Toolbox, page 782)*

24. $(2x - 3)(2x + 3)$

25. $(m + 5)^2$

26. $(2n - 4)(3n + 1)$

Section 9.2

Multiplying and Dividing Polynomials

Learn how to . . .

- multiply and divide polynomials

So you can . . .

- solve problems that involve products or quotients of polynomials

Application

A baseball player's metal spikes exert a great amount of pressure on a solid floor, as described by the expression

$$\frac{\text{weight of the player}}{\text{total base area of the spikes}}.$$

When the weight of the player is expressed as a function of height and the total base area of the spikes is expressed as a function of the base area of each spike and the number of spikes, you can divide polynomials to calculate the pressure exerted.

UNDERSTANDING THE MAIN IDEAS

Multiplying polynomials

You can use the distributive property to multiply two polynomials. When using the distributive property, you multiply each term of one polynomial by each term of the other polynomial. A vertical or horizontal format can be used to write out this multiplication.

Example 1

Find each product.

a. $(5x^2 - 3x - 7)(3x - 2)$ **b.** $(2y^3 + 5y - 3)(4y^3 + 1)$

Solution

a. *Method 1*: Use a horizontal format.

To use the horizontal format, multiply the first factor by $3x$, and then multiply the first factor by -2.

$$(5x^2 - 3x - 7)(3x - 2) = (5x^2 - 3x - 7)(3x) + (5x^2 - 3x - 7)(-2)$$
$$= (15x^3 - 9x^2 - 21x) + (-10x^2 + 6x + 14)$$
$$= 15x^3 - 19x^2 - 15x + 14$$

(Solution continues on next page.)

■ **Solution** ■ *(continued)*

Method 2: Use a vertical format.

$$5x^2 - 3x - 7$$
$$\underline{3x - 2}$$
$$\underline{15x^3 - 9x^2 - 21x} \qquad \leftarrow 3x(5x^2 - 3x - 7)$$
$$\underline{ - 10x^2 + 6x + 14} \qquad \leftarrow -2(5x^2 - 3x - 7)$$
$$15x^3 - 19x^2 - 15x + 14$$

The product is $15x^3 - 19x^2 - 15x + 14$.

b. Method 1: $(2y^3 + 5y - 3)(4y^3 + 1) = (2y^3 + 5y - 3)(4y^3) + (2y^3 + 5y - 3)(1)$
$$= 8y^6 + 20y^4 - 12y^3 + 2y^3 + 5y - 3$$
$$= 8y^6 + 20y^4 - 10y^3 + 5y - 3$$

Method 2: Use $0y^2$ for the "missing" y^2-term in the first factor.

$$2y^3 + 0y^2 + 5y - 3$$
$$\underline{4y^3 + 1}$$
$$\underline{8y^6 + 0y^5 + 20y^4 - 12y^3}$$
$$\underline{ 2y^3 + 0y^2 + 5y - 3}$$
$$8y^6 + 20y^4 - 10y^3 + 5y - 3$$

The product is $8y^6 + 20y^4 - 10y^3 + 5y - 3$.

Multiply.

1. $(2x - 3)(4x + 7)$
2. $3y^2(4y^3 - 8y^2 + y - 7)$
3. $(-6x^3 - x + 2)(-2x^2 - 5)$

4. $(3s + 7)(s^2 - s - 1)$
5. $(3x - 2)^3$
6. $(2y - 1)^4$

Dividing polynomials

To divide polynomials, you can use a process similar to the long division you learned in elementary school.

Example 2

Find each quotient.

a. $\dfrac{4x^3 - 16x^2 + 19x - 10}{2x - 5}$

b. $\dfrac{25y^3 - 4y + 13}{5y + 3}$

■ Solution ■

a. Use the first term of the denominator, $2x$, to determine each successive term of the quotient.

$$
\begin{array}{r}
2x^2 - 3x + 2 \\
2x - 5\overline{)4x^3 - 16x^2 + 19x - 10} \\
\underline{4x^3 - 10x^2} \qquad\qquad \leftarrow \text{Since } 4x^3 \div 2x = 2x^2, \text{ multiply } 2x - 5 \text{ by } 2x^2. \\
-6x^2 + 19x \\
\underline{-6x^2 + 15x} \qquad\qquad \leftarrow \text{Since } -6x^2 \div 2x = -3x, \text{ multiply } 2x - 5 \text{ by } -3x. \\
4x - 10 \\
\underline{4x - 10} \leftarrow \text{Since } 4x \div 2x = 2, \text{ multiply } 2x - 5 \text{ by } 2. \\
0
\end{array}
$$

Therefore, $\dfrac{4x^3 - 16x^2 + 19x - 10}{2x - 5} = 2x^2 - 3x + 2$.

b. When there are missing terms in the given numerator, insert that term with a coefficient of 0 before beginning the long division. In this case, insert the term $0y^2$ in the dividend as shown below.

$$
\begin{array}{r}
5y^2 - 3y + 1 \\
5y + 3\overline{)25y^3 + 0y^2 - 4y + 13} \\
\underline{25y^3 + 15y^2} \qquad\qquad \leftarrow \text{Since } 25y^3 \div 5y = 5y^2, \text{ multiply } 5y + 3 \text{ by } 5y^2. \\
-15y^2 - 4y \\
\underline{-15y^2 - 9y} \qquad\qquad \leftarrow \text{Since } -15y^2 \div 5y = -3y, \text{ multiply } 5y + 3 \text{ by } -3y. \\
5y + 13 \\
\underline{5y + 3} \leftarrow \text{Since } 5y \div 5y = 1, \text{ multiply } 5y + 3 \text{ by } 1. \\
10
\end{array}
$$

There is a remainder for this division. (*Note*: You stop the long division process when either the degree of the remainder is less than the degree of the divisor, or when the remainder is 0.)

Therefore, $\dfrac{25y^3 - 4y + 13}{5y + 3} = 5y^2 - 3y + 1 + \dfrac{10}{5y + 3}$. (Notice how the remainder is written as the numerator of the final term with the divisor as the denominator.)

For Exercises 7–12, divide.

7. $\dfrac{8x^2 + 2x - 21}{2x - 3}$

8. $\dfrac{8x^2 + 2x - 21}{4x + 7}$

9. $\dfrac{3s^3 + 4s^2 - 10s - 7}{3s + 7}$

10. $\dfrac{3s^3 + 4s^2 - 10s - 5}{3s + 7}$

11. $\dfrac{27x^3 - 54x^2 + 36x - 8}{3x - 2}$

12. $\dfrac{4x^4 - 9}{2x^2 + 3}$

13. Mathematics Journal Look closely at Exercises 7 and 8. Describe their relationship. Compare Exercises 4 and 9. How are they related? Are there any other pairs of exercises that are related to each other? If so, describe the relationships.

Use synthetic substitution to evaluate each polynomial for the given value of the variable. *(Section 9.1)*

14. $5y^3 - 4y^2 + 2y - 1; y = 3$

15. $\frac{1}{2}x^2 - 4x - 5; x = 6$

TECHNOLOGY **Use a graphing calculator or software with matrix calculation capabilities to solve each system of equations.** *(Section 7.3)*

16. $-20x + 30y = -60$
$-15x - 12y = 93$

17. $2x - 3y + z = 11$
$4x + y + 2z = 15$
$x - 3y - 5z = -15$

Graph each equation. *(Sections 2.2 and 5.4)*

18. $y = -\frac{3}{4}x - 4$

19. $y = -2x^2 + 7x - 5$

Exploring Graphs of Polynomial Functions

Learn how to ...

- recognize graphs of polynomial functions and describe their important features

So you can ...

- use graphs to solve problems

Application

A swimmer's speed doing the backstroke changes several times during each stroke. You can use the graph of an equation that relates speed and time to identify when, during the stroke, the swimmer's speed reaches its maximum and minimum values. The equation is a 7th-degree polynomial, where s represents the speed in meters per second and t represents the time in seconds.

$$s = -241.0t^7 + 1062t^6 - 1871t^5 + 1647t^4 - 737.4t^3 + 143.9t^2 - 2.432t$$

Terms to Know	Example / Illustration
End behavior (p. 405) what happens to a function $f(x)$ as x takes on large positive or negative values (There are four types of end behavior for polynomial functions.)	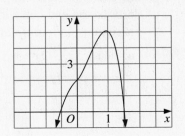 As $x \to +\infty$, $f(x) \to -\infty$. As $x \to -\infty$, $f(x) \to -\infty$.
Leading coefficient (p. 406) the coefficient of the highest power of the variable in a polynomial function	The leading coefficient of the function $f(x) = -3x^4 + 2x^3 + x^2 + 3x + 2$ is -3.
Turning point (p. 407) a point on the graph of a function that is higher or lower than all nearby points (In general, the graph of a polynomial function of degree n has at most $n - 1$ turning points.)	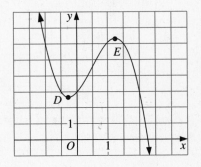 Points D and E are turning points of the graph of $f(x) = -2x^3 + 3x^2 + 2x + 3$.

Local maximum (minimum) (p. 407) the *y*-coordinate of a turning point that is higher (lower) than all nearby points.	In the graph shown above, the *y*-coordinate of point *D* is a local minimum and the *y*-coordinate of point *E* is a local maximum.

UNDERSTANDING THE MAIN IDEAS

Describing graphs

For a polynomial function, you can describe its end behavior by looking at the exponent and coefficient of its leading term. You can also identify any local maximums and minimums from its graph.

Example

For each function and graph, relate its end behavior to its leading term, and describe the end behavior using infinity notation. Also, identify any local maximums and minimums.

a. $f(x) = 3x^5 - 2x^4 + 5x - 3$

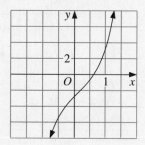

b. $g(x) = -5x^4 + 3x^3 - x^2 + x + 2$

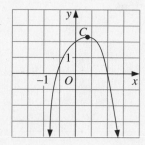

c. $h(x) = 2x^6 + 3x^3 + 5$

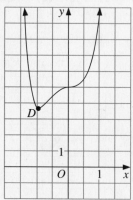

d. $k(x) = -2x^7 - 5x^6 + 4x^2 + 7$

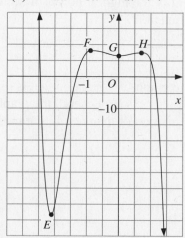

■ Solution ■

In general, if the exponent of the leading term is even, both ends of the graph go in the same direction; if the exponent of the leading term is odd, the ends of the graph go in opposite directions.

a. The exponent of the leading term, $3x^5$, is odd so the ends of the graph go in opposite directions.

> As $x \rightarrow +\infty$, $f(x) \rightarrow +\infty$.
> As $x \rightarrow -\infty$, $f(x) \rightarrow -\infty$.

There are no local maximums or minimums.

b. The exponent of the leading term, $-5x^4$, is even so the ends of the graph go in the same direction. The coefficient is negative, so both ends of the graph go to $-\infty$.

> As $x \rightarrow +\infty$, $g(x) \rightarrow -\infty$.
> As $x \rightarrow -\infty$, $g(x) \rightarrow -\infty$.

The graph has one turning point, labeled C, which is a local maximum.

c. The exponent of the leading term, $2x^6$, is even so the ends of the graph go in the same direction. The coefficient is positive, so both ends of the graph go to $+\infty$.

> As $x \rightarrow +\infty$, $h(x) \rightarrow +\infty$.
> As $x \rightarrow -\infty$, $h(x) \rightarrow +\infty$.

The graph has one turning point, labeled D. Point D is a local minimum.

d. The exponent of the leading term, $-2x^7$, is odd so the ends of the graph go in opposite directions.

> As $x \rightarrow +\infty$, $k(x) \rightarrow -\infty$.
> As $x \rightarrow -\infty$, $k(x) \rightarrow +\infty$.

The graph has four turning points. Points E and G are local minimums and points F and H are local maximums.

TECHNOLOGY Use a graphing calculator or software to graph each polynomial function. For each function:

a. **Describe the end behavior using infinity notation.**

b. **Find all local maximums and minimums.**

1. $f(x) = -3x^4 + 2x^3 - 5x$ **2.** $f(x) = 2x^6 - 5x^5 - 3x^2 - 5$

3. $g(x) = -x^3 + 2x^2 - 3x - 5$ **4.** $g(x) = -2x^8 + 5x^5 - x^4 + 3$

5. $f(x) = 5x^9 - 3x^8 + 2x^5 - 3$ **6.** $f(x) = x^8 - x^7 + x^6 - x^5 + 4$

7. $g(x) = -7x^3 + 2x^2 - 3x + 4$ **8.** $g(x) = 5x^5 + 4x^4 + 3x^3 + 2x^2 + x$

For each function:

a. Classify the exponent of the leading term as even or odd, and classify the coefficient of the leading term as positive or negative.

b. Use the information from part (a) to describe the end behavior of the function.

9. $f(x) = -5x^6 + 4x^5 + 2x - 3$

10. $f(x) = -7x^9 - 2x + 3$

11. $g(x) = -3 + 2x + 5x^5$

12. $g(x) = -7x + 3 + 5x^4$

For Exercises 13–16, the end behavior of a polynomial function is given. State whether the exponent of the leading term of each function is *even* or *odd*, and whether the coefficient of the leading term is *positive* or *negative*.

13. As $x \to +\infty$, $f(x) \to -\infty$.
As $x \to -\infty$, $f(x) \to +\infty$.

14. As $x \to +\infty$, $f(x) \to +\infty$.
As $x \to -\infty$, $f(x) \to +\infty$.

15. As $x \to +\infty$, $g(x) \to -\infty$.
As $x \to -\infty$, $g(x) \to -\infty$.

16. As $x \to +\infty$, $g(x) \to +\infty$.
As $x \to -\infty$, $g(x) \to -\infty$.

....
Spiral Review

Write a point-slope equation of the line passing through the given points.
(Section 2.3)

17. $(1, -2)$ and $(-3, -14)$

18. $(9, 3)$ and $(-3, -5)$

19. $(-15, 8)$ and $(0, 2)$

Solve each equation. *(Sections 5.5 and 5.6)*

20. $2x^2 - 13x + 18 = 0$

21. $x^2 + 4x - 6 = 0$

22. $x^2 = 3x + 11$

Solving Cubic Equations

Learn how to . . .

- solve cubic
  equations

- find equations for
  graphs of cubic
  functions

- find zeros of cubic
  functions

So you can . . .

- solve problems
  involving cubic
  functions

Application

The tachometer of a motorboat shows the number of revolutions per minute made by the motorboat's propeller. The speed of the motorboat is a cubic function of the tachometer rate.

$$s = 0.00547r^3 - 0.224r^2 + 3.60r - 11.0$$

Terms to Know	*Example / Illustration*
Cubic function (p. 413) a polynomial function of degree 3	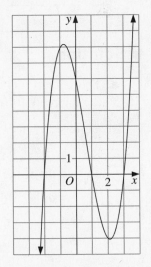 The graph of the cubic function $f(x) = x^3 - 2x^2 - 5x + 6$ is shown above.
Zero (p. 413) a solution of the equation $f(x) = 0$, where $f(x)$ is a polynomial function	The zeros of $f(x) = x^3 - 2x^2 - 5x + 6$ are $x = -2$, 1, and 3. That is, $f(-2) = 0$, $f(1) = 0$, and $f(3) = 0$.

UNDERSTANDING THE MAIN IDEAS

Zeros and intercepts

The real-number zeros of a polynomial function correspond to the x-intercepts of its graph. Cubic functions may be written in *standard form*, $f(x) = ax^3 + bx^2 + cx + d$, or in *intercept form*, $f(x) = a(x - p)(x - q)(x - r)$, where p, q, and r are the x-intercepts of the graph of the function.

Example 1

For each graph, write a cubic function in intercept form. Then state the zeros of the function.

a.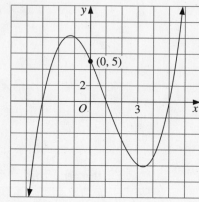

b.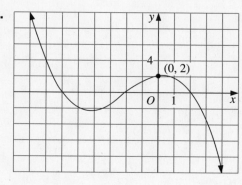

■ Solution ■

a. The x-intercepts of the graph are -3, 1, and 5. So let $p = -3$, $q = 1$, and $r = 5$ in the intercept form of a cubic function.

$$f(x) = a(x - (-3))(x - 1)(x - 5)$$
$$= a(x + 3)(x - 1)(x - 5)$$

To find the value of a, use the point $(0, 5)$, which is on the graph of f.

$$5 = a(0 + 3)(0 - 1)(0 - 5) \quad \leftarrow x = 0, y = 5$$
$$5 = a(3)(-1)(-5)$$
$$5 = 15a$$
$$a = \frac{1}{3}$$

An equation of the function is $f(x) = \frac{1}{3}(x + 3)(x - 1)(x - 5)$. The zeros of the function are -3, 1, and 5.

(Solution continues on next page.)

■ **Solution** ■ (*continued*)

    b. The *x*-intercepts of the graph are –6, –2, and 2. So let $p = -6$, $q = -2$, and $r = 2$ in the intercept form of a cubic function.

$$f(x) = a(x - (-6))(x - (-2))(x - 2)$$
$$= a(x + 6)(x + 2)(x - 2)$$

    To find the value of *a*, use the point (0, 2), which is on the graph of *f*.

$$2 = a(0 + 6)(0 + 2)(0 - 2) \quad \leftarrow x = 0, y = 2$$
$$2 = a(6)(2)(-2)$$
$$2 = -24a$$
$$a = -\frac{1}{12}$$

    An equation of the function is $f(x) = -\frac{1}{12}(x + 6)(x + 2)(x - 2)$. The zeros of the function are –6, –2, and 2.

State the zeros of each function.

1. $f(x) = 3(x + 5)(x + 1)(x - 7)$

2. $f(t) = (2t + 1)(3t + 1)(2t - 5)$

3. $g(n) = \frac{1}{2}\left(n + \frac{1}{2}\right)\left(n - \frac{1}{3}\right)\left(n + \frac{1}{4}\right)$

4. $g(x) = \left(x + \frac{2}{3}\right)\left(x - \frac{1}{2}\right)(x - 2)$

Find an equation for each cubic function whose graph is shown.

5.
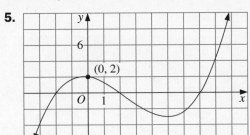

6.
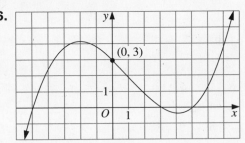

Factors and zeros

There are two important properties of polynomials. The first is called the *factor theorem*:

    For any polynomial $f(x)$, $x - k$ is a factor of $f(x)$ if and only if $f(k) = 0$.

When finding the zeros of a cubic function $f(x)$ given in standard form, the factor theorem says that if you can find one zero *k*, then you can write $f(x)$ as

$$f(x) = (x - k) \cdot q(x)$$

where $q(x) = \frac{f(x)}{x - k}$ is a polynomial of degree 2. You can then solve the quadratic

equation $q(x) = 0$ using one of the techniques you learned in Chapter 5. The resulting zeros of *q* are also zeros of *f*.

The second property holds for any polynomial $f(x)$ with coefficients that are integers:

The possible integral zeros of f are the divisors of the constant term.

(*Note*: The property above does not guarantee any integral zeros, only that if there are integral zeros they must be divisors of the constant term.)

You can use these two properties, along with synthetic substitution, to find all the zeros of a cubic function.

Example 2

Find the zeros of each function.

a. $f(x) = 2x^3 - 11x^2 - x + 30$ **b.** $h(x) = x^3 + 7x^2 + 8x + 2$

■ Solution ■

a. All the coefficients are integers, so the possible *integral* zeros of f are the divisors of the constant term, 30. They are:

$\pm 1, \pm 2, \pm 3, \pm 5, \pm 6, \pm 10, \pm 15, \pm 30$.

Use synthetic substitution to find one of the zeros; that is, to find a value of x so that $f(x) = 0$. Recall that the last sum found using synthetic substitution is the value of the function for the given value of x.

Try $x = -1$:

$$\begin{array}{r|rrrr} -1 & 2 & -11 & -1 & 30 \\ & & -2 & 13 & -12 \\ \hline & 2 & -13 & 12 & 18 \end{array}$$

Try $x = 1$:

$$\begin{array}{r|rrrr} 1 & 2 & -11 & -1 & 30 \\ & & 2 & -9 & -10 \\ \hline & 2 & -9 & -10 & 20 \end{array}$$

Try $x = -2$:

$$\begin{array}{r|rrrr} -2 & 2 & -11 & -1 & 30 \\ & & -4 & 30 & 58 \\ \hline & 2 & -15 & -29 & 88 \end{array}$$

Try $x = 2$:

$$\begin{array}{r|rrrr} 2 & 2 & -11 & -1 & 30 \\ & & 4 & -14 & -30 \\ \hline & 2 & -7 & -15 & 0 \end{array}$$

$f(2) = 0$, so 2 is a zero.

Now you know that $x = 2$ is a zero of $f(x)$. So by the factor theorem, $x - 2$ is a factor of $f(x)$. You could use long division now to find $q(x)$ such that $(x - 2) \cdot q(x) = f(x)$. However, one of the advantages of using synthetic substitution is that when you find a zero, the numbers in the bottom row preceding the final zero are the coefficients of the function $q(x)$. In this case, $q(x) = 2x^2 - 7x - 15$.

$$\begin{aligned} f(x) &= 2x^3 - 11x^2 - x + 30 \\ &= (x - 2)(2x^2 - 7x - 15) \qquad \leftarrow \text{Factor } 2x^2 - 7x - 15 \text{ if possible.} \\ &= (x - 2)(2x + 3)(x - 5) \end{aligned}$$

Therefore, the zeros of $f(x)$ are 2, $-\dfrac{3}{2}$, and 5.

(Solution continues on next page.)

■ **Solution** ■ *(continued)*

b. All the coefficients are integers, so the possible zeros of $h(x)$ are the divisors of 2: $\pm 1, \pm 2$.

Try $x = 1$:

$$\begin{array}{r|rrr} 1 & 1 & 7 & 8 & 2 \\ & & 1 & 8 & 16 \\ \hline & 1 & 8 & 16 & 18 \end{array}$$

Try $x = -1$:

$$\begin{array}{r|rrr} -1 & 1 & 7 & 8 & 2 \\ & & -1 & -6 & -2 \\ \hline & 1 & 6 & 2 & 0 \end{array} \quad h(-1) = 0$$

So $x = -1$ is a zero and $q(x) =$ and $x^2 + 6x + 2$. Therefore,

$$h(x) = x^3 + 7x^2 + 8x + 2$$
$$= (x - (-1))(x^2 + 6x + 2)$$
$$= (x + 1)(x^2 + 6x + 2)$$

But $x^2 + 6x + 2$ is not factorable, so use the quadratic formula to find the remaining zeros of h.

Let $x^2 + 6x + 2 = 0$:

$$x = \frac{-6 \pm \sqrt{6^2 - 4(1)(2)}}{2(1)}$$
$$= \frac{-6 \pm \sqrt{36 - 8}}{2}$$
$$= \frac{-6 \pm 2\sqrt{7}}{2}$$
$$= -3 \pm \sqrt{7}$$

Therefore, the zeros of $h(x)$ are -1, $-3 - \sqrt{7}$, and $-3 + \sqrt{7}$.

Find the zeros of each function.

7. $f(x) = x^3 + 2x^2 - 11x - 12$

8. $h(x) = x^3 - 39x + 70$

9. $f(x) = 15x^3 - 22x^2 + 5x + 2$

10. $h(x) = x^3 + 2x^2 - 11x + 8$

....
Spiral Review

Describe the end behavior of each function using infinity notation. *(Section 9.3)*

11. $f(x) = 15x^5 - 3x^4 + 2x - 3$

12. $f(x) = 2x^6 - x^3 + 15x$

13. $g(x) = -x^4 - x^3 + x^2 + x - 3$

14. $g(x) = -2x^3 - 3x^2 + 4x - 1$

Simplify. *(Section 3.2)*

15. 7^{-3}

16. $16^{5/2}$

17. $49^{3/2}$

Solve using the quadratic formula. Check your solutions. *(Section 8.4)*

18. $x^2 - 3 = 4x$

19. $3x^2 = 2x + 6$

Finding Zeros of Polynomial Functions

Learn how to . . .

- find zeros of higher-degree polynomial functions

So you can . . .

- solve problems involving higher-degree polynomial functions

Application

When a person stands on a diving board, the bend or *deflection* of the diving board is a function of where the person is standing on the board. This deflection can be modeled by a polynomial function.

Terms to Know

Example / Illustration

Double zero (p. 420) for a polynomial function $f(x)$, the number k if $(x - k)^2$ is a factor of f	If $f(x) = 5(x + 3)(x - 5)^2$, then 5 is a double zero of $f(x)$.
Triple zero (p. 420) for a polynomial function $f(x)$, the number k if $(x - k)^3$ is a factor of f	If $g(x) = (x + 3)(x + 1)^3(x - 4)^3$, then -1 and 4 are both triple zeros of $g(x)$.

UNDERSTANDING THE MAIN IDEAS

Number of complex zeros

While a polynomial function of degree n has at most n real zeros, the fundamental theorem of algebra states that a polynomial function of degree n has exactly n complex zeros, provided each double zero is counted as 2 zeros, each triple zero is counted as 3 zeros, and so on. For example, a cubic function can have 1, 2, or 3 real zeros. Since it has exactly 3 complex zeros, if only one of them is a real zero then the other two are imaginary zeros. When there are two real zeros of a cubic function, then one of them must be a double zero.

Example 1

Tell the number of real and the number of imaginary zeros for each polynomial function. Then find each complex zero.

a. $f(x) = 5(x - 3)^2(x + 4)^3(x - 5)$ **b.** $g(x) = (x^2 - 5x + 6)(x^2 + 4)$

▪ Solution ▪

a. The degree of the function is 6, so the fundamental theorem of algebra indicates there are exactly six complex zeros. The function has a double zero at $x = 3$, a triple zero at $x = -4$, and a zero at $x = 5$. Since these six zeros are real, there are no imaginary zeros.

b. The degree of the function is 4, so there are exactly four complex zeros. To find the zeros, solve these equations:

$$x^2 - 5x + 6 = 0 \qquad \text{or} \qquad x^2 + 4 = 0$$
$$(x - 3)(x - 2) = 0 \qquad \text{or} \qquad x^2 = -4$$
$$x - 3 = 0 \quad \text{or} \quad x - 2 = 0 \qquad \text{or} \qquad x = \pm\sqrt{-4}$$
$$x = 3 \quad \text{or} \quad x = 2 \qquad \text{or} \qquad x = \pm 2i$$

There are two real zeros, 2 and 3, and two imaginary zeros, $-2i$ and $2i$.

Find all real and imaginary zeros of each function. Identify any double or triple zeros.

1. $f(x) = 6(x - 3)^3(x + 2)^3(x - 4)$ **2.** $f(x) = (x^2 - 6x + 5)(x^2 - 5x + 4)$

3. $g(x) = (x^2 + x + 1)(2x^2 + 3x + 1)$ **4.** $g(x) = 2x^3 - 5x^2 + 7x - 4$

Rational zeros of polynomial functions

Many polynomial functions have rational numbers as zeros. A rational zero is a zero of the form $\frac{p}{q}$, where p and q are integers. The *rational zeros theorem* states that if f is a polynomial function with integral coefficients, then the *only* possible rational zeros of f are $\frac{p}{q}$, where p is a divisor of the constant term of $f(x)$ and q is a divisor of the leading coefficient. For example, suppose the leading coefficient and the constant term of a polynomial function with integral coefficients are 2 and 6, respectively. Since the divisors of 2 are ± 1 and ± 2, and the divisors of 6 are ± 1, ± 2, ± 3, and ± 6, the possible rational zeros of the function are $\pm\frac{1}{1} = \pm 1$, $\pm\frac{2}{1} = \pm 2$, $\pm\frac{3}{1} = \pm 3$, $\pm\frac{6}{1} = \pm 6$, $\pm\frac{1}{2}$, and $\pm\frac{3}{2}$.
(Notice that the possible rational zeros include all the possible real zeros.)

Example 2

Find all the zeros of the polynomial function $f(x) = 9x^4 - 12x^3 + 13x^2 - 12x + 4$.

■ Solution ■

The fundamental theorem of algebra indicates that $f(x)$ has exactly four complex zeros. Notice that $f(x)$ is a polynomial function with integral coefficients.

Step 1: Find the possible rational zeros of f.

divisors of the constant term 4: $\pm 1, \pm 2, \pm 4$

divisors of the leading coefficient 9: $\pm 1, \pm 3, \pm 9$

So the possible rational zeros are $\pm 1, \pm 2, \pm 4, \pm\frac{1}{3}, \pm\frac{2}{3}, \pm\frac{4}{3}, \pm\frac{1}{9}, \pm\frac{2}{9},$ and $\pm\frac{4}{9}$.

Step 2: Use synthetic substitution to find one zero of f from the list in Step 1. (While you must search the list until you find a zero, only the successful outcome is shown here.)

Try $x = \frac{2}{3}$:

$$
\begin{array}{r|rrrrr}
\frac{2}{3} & 9 & -12 & 13 & -12 & 4 \\
 & & 6 & -4 & 6 & -4 \\
\hline
 & 9 & -6 & 9 & -6 & 0 \\
\end{array}
$$

So $x = \frac{2}{3}$ is a zero of f and $f(x) = \left(x - \frac{2}{3}\right)(9x^3 - 6x^2 + 9x - 6)$.

Step 3: Now find another zero of f by finding a zero of $q(x) = 9x^3 - 6x^2 + 9x - 6$. (When searching for a second zero, use the list from Step 1 but do not start at the beginning of the list again. Begin with the successful zero found in Step 2 since it might be a double zero. Again, only the successful outcome is shown here.)

$$
\begin{array}{r|rrrr}
\frac{2}{3} & 9 & -6 & 9 & -6 \\
 & & 6 & 0 & 6 \\
\hline
 & 9 & 0 & 9 & 0 \\
\end{array}
$$

This shows that $x = \frac{2}{3}$ is a double zero, and that $f(x) = \left(x - \frac{2}{3}\right)^2(9x^2 + 9)$.

Step 4: Solve $9x^2 + 9 = 0$ to find the other two zeros.

$$9x^2 + 9 = 0$$
$$9x^2 = -9$$
$$x^2 = -1$$
$$x = \pm\sqrt{-1}$$
$$= \pm i$$

The zeros of f are $x = \frac{2}{3}$ (a double zero), $x = i$, and $x = -i$.

List the possible rational zeros for each function.

5. $f(x) = 80x^3 - 32x^2 - 5x + 2$

6. $f(x) = 24x^3 + 26x^2 + 9x + 1$

7. $f(x) = 16x^4 - 40x^3 - 31x^2 + 22x - 3$

8. $f(x) = 8x^4 + 20x^3 - 18x^2 - 81x - 54$

Find all real and imaginary zeros of each function. Identify any double or triple zeros.

9. $f(x) = 80x^3 - 32x^2 - 5x + 2$

10. $f(x) = 24x^3 + 26x^2 + 9x + 1$

11. $f(x) = 16x^4 - 40x^3 - 31x^2 + 22x - 3$

12. $f(x) = 8x^4 + 20x^3 - 18x^2 - 81x - 54$

Refer to the Application at the beginning of this section. The deflection d, in inches, of a diving board for a person who weighs 150 lb is given by the function

$$d(x) = (-4.752 \times 10^{-7})x^3 + (1.711 \times 10^{-4})x^2$$

when the person stands x inches from the supported end of the board. Use this information for Exercises 13 and 14.

13. Use a calculator to find the deflection of the board when the person is 60 in. from the supported end.

14. What is the change in deflection when the person moves from 5 ft from the supported end of the board to 15 ft from the supported end of the board?

15. Open-ended Write three binomials of the form $ax + b$, where a and b are constants. If the product of your three binomials is zero, do you have a cubic equation? What are the zeros of your equation?

· · · · · · · · · · · · · · · · · ·
Spiral Review

Find an equation for the cubic function whose graph has the given x-intercepts and passes through the given point. *(Section 9.4)*

16. x-intercepts: $-1, 3, 5$; $(0, 15)$

17. x-intercepts: $-4, -2, 2$; $(0, -4)$

For each equation, tell whether y varies directly with x. *(Section 2.1)*

18. $y = \dfrac{2}{3}x$

19. $y = -x$

20. $y = x + 5$

21. $y = \dfrac{1}{3x}$

Inverse Variation

Learn how to . . .

- recognize inverse variation

- write and use inverse variation equations

So you can . . .

- solve problems involving inverse variation

Application

A plumber's wrenches and pliers have handles of various lengths. The longer handles require less force to turn a bolt or pipe. The length of the handle and the applied force show inverse variation; that is, they are inversely proportional.

Terms to Know

Example / Illustration

Inverse variation (p. 427) two variables x and y that are related by an equation of the form $y = \dfrac{a}{x}$	The equation $y = \dfrac{2}{x}$ is an example of inverse variation. (The inverse variation equation can also be written as $xy = 2$.)
Constant of variation (p. 427) the value of a $(a \neq 0)$ in the equation $y = \dfrac{a}{x}$	In the inverse variation equation above, the constant of variation is 2.
Hyperbola (p. 427) the graph of an inverse variation equation (A hyperbola consists of two branches.)	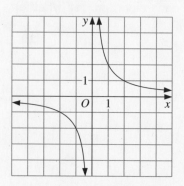 The graph of $y = \dfrac{2}{x}$ is a hyperbola, which has two branches.

Terms to Know	Example / Illustration
Asymptote (p. 427) a line that a graph approaches more and more closely	In the graph on the previous page, the two branches of the hyperbola both approach the *x*-axis and the *y*-axis, which are the asymptotes of the graph.

UNDERSTANDING THE MAIN IDEAS

Equations and graphs for inverse variation

If the values of one of two related variables get larger as the values of the other variable get smaller, the variables may show inverse variation. Another way to state an inverse variation relationship is "two variables *x* and *y* vary inversely if and only if the product *xy* is a constant." This constant is the constant of variation.

Example

For each equation, tell whether *y* varies inversely with *x*. If so, state the constant of variation and graph the equation.

a. $y = \dfrac{-4}{x}$ **b.** $y = -4x$ **c.** $y = -4x^2$

Solution

a. The equation $y = \dfrac{-4}{x}$ has the form $y = \dfrac{a}{x}$, so it represents inverse variation and *y* varies inversely with *x*. The constant of variation is –4. The graph, a hyperbola with branches located in Quadrants II and IV with the *x*- and *y*-axes as asymptotes, is shown below.

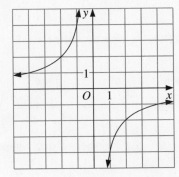

(Solution continues on next page.)

---

b. The equation $y = -4x$ has the form $y = ax$, so it does not represent inverse variation and y does not vary inversely with x. (The equation $y = -4x$ is an example of direct variation, which you studied in Section 2.1.)

c. The equation $y = -4x^2$ does not represent inverse variation and y does not vary inversely with x. (The equation $y = -4x^2$ is that of a parabola, which you studied in Sections 5.1 and 5.2.)

Write each equation in the form $y = \dfrac{a}{x}$.

1. $xy = 10$

2. $-6xy = 3$

3. $4y = \dfrac{20}{x}$

4. $x = \dfrac{16}{y}$

Each group of ordered pairs represents solutions of an inverse variation equation. Write the inverse variation equation in the form $y = \dfrac{a}{x}$ and identify the constant of variation.

5. $(1, 3), \left(6, \dfrac{1}{2}\right)$

6. $\left(-20, \dfrac{1}{5}\right), (-4, 1), (-2, 2)$

7. $\left(\dfrac{1}{3}, -3\right), \left(7, -\dfrac{1}{7}\right), (1, -1)$

8. $(1, m), \left(6, \dfrac{1}{6}m\right)$

Tell whether the given values of x and y show inverse variation. If so, write an equation giving y as a function of x.

9.

x	y
2	-1
4	$-\dfrac{1}{2}$
6	$-\dfrac{1}{3}$
8	$-\dfrac{1}{4}$

10.

x	y
2	-4
4	-8
6	-12
8	-16

11.

x	y
2	1
4	2
6	3
8	4

12.

x	y
2	8
4	4
6	$\dfrac{8}{3}$
8	2

Tell whether you think the two quantities show inverse variation. Explain your reasoning.

13. the number of equal pieces of a pie and the size of each piece

14. the number of tickets sold at a raffle and your chance to win the raffle prize by buying one ticket

15. the number of pounds of ground beef and the number of quarter-pound hamburger patties that can be made

16. the speed of a plane and the amount of time needed to fly between two cities

The force F needed to turn a pipe using a pipe wrench varies inversely with the length l of the pipe wrench. Suppose a pipe wrench 10 in. long required a plumber to exert a force of 75 lb in order to turn a certain pipe fitting. Use this information for Exercises 17–19.

17. Write an equation relating F and l.

18. What force would have been needed if the plumber had used a similar pipe wrench 25 in. long to turn the same pipe fitting?

19. Suppose the pipe was in a corner, and the only pipe wrench that the plumber could use in the confined space was 6 in. long. What force would have been necessary to turn the pipe fitting?

20. Mathematics Journal Write about any tools or other instruments you have used that take advantage of the length of a handle.

....
Spiral Review

Find all real and imaginary zeros of each function. Identify any double or triple zeros. *(Section 9.5)*

21. $f(x) = x^3 - 9x^2 + 24x - 20$ **22.** $f(x) = 8x^4 - 12x^3 - 30x^2 - 17x - 3$

Describe how the graph of each equation is related to the graph of $y = 3x^2$. *(Section 5.2)*

23. $y = 3x^2 - 4$ **24.** $y = 3(x + 2)^2$ **25.** $y = -3x^2$

Working with Simple Rational Functions

Learn how to . . .

- identify important features of translated hyperbolas
- find equations of translated hyperbolas

So you can . . .

- use translations of hyperbolas to solve problems

Application

When you see lightning strike the ground, a well-known way to approximate your distance in miles from the point where it struck is by counting the seconds until you hear thunder, and dividing by 5. A more accurate relationship is given by the expression

$$t = \frac{d}{1.09T + 1050}$$

where t is the time (in seconds), d is your distance from the lightning (in feet), and T is the air temperature (in degrees Fahrenheit). The expression represents a rational function.

Terms to Know

Example / Illustration

Rational function (p. 435)

a function of the form $f(x) = \frac{p(x)}{q(x)}$, where $p(x)$ and $q(x)$ are polynomials

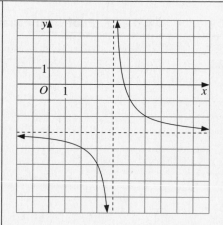

The graph of $f(x) = \frac{-3x + 14}{x - 4}$ above is a hyperbola with vertical asymptote $x = 4$ and horizontal asymptote $y = -3$.

Equations and graphs

If you can rewrite a rational function in the form $f(x) = \dfrac{a}{x-h} + k$, then you can

describe its graph in terms of the hyperbola for the equation $y = \dfrac{a}{x}$. The graph of

$f(x) = \dfrac{a}{x-h} + k$ is a hyperbola that is the translation of the graph of $y = \dfrac{a}{x}$

horizontally h units and vertically k units. Furthermore, the asymptotes of the

graph of $f(x) = \dfrac{a}{x-h} + k$ are $y = k$ and $x = h$. The vertical line $x = h$ is an

asymptote because $f(x)$ does not have a value at $x = h$ since it makes the value of

the denominator in the function 0. Also, since the value of the expression $\dfrac{a}{x-h}$

cannot be 0 (remember, a is a nonzero constant), then the value of $f(x)$ cannot be

k and the line $y = k$ is a horizontal asymptote.

Example 1

Tell whether each function is a rational function. Then describe the graph of the function.

a. $f(x) = \dfrac{3}{x-2} + 5$ **b.** $g(x) = \dfrac{-1}{x+4} - 2$

■ Solution ■

a. Rewrite the function.

$$f(x) = \frac{3}{x-2} + 5$$

$$= \frac{3}{x-2} + \frac{5(x-2)}{x-2}$$

$$= \frac{3 + 5x - 10}{x-2}$$

$$= \frac{5x - 7}{x-2}$$

Since $f(x)$ has the form $\dfrac{p(x)}{q(x)}$, where $p(x)$ and $q(x)$ are polynomials, $f(x)$ is a rational function.

The graph of the function $f(x) = \dfrac{3}{x-2} + 5$, is a translation of the graph of $y = \dfrac{3}{x}$ to the right 2 units ($h = 2$) and up 5 units ($k = 5$). The graph has a vertical asymptote at $x = 2$ and a horizontal asymptote at $y = 5$.

(Solution continues on next page.)

b. $g(x) = \dfrac{-1}{x+4} - 2$

$\qquad = \dfrac{-1}{x+4} - \dfrac{2(x+4)}{x+4}$

$\qquad = \dfrac{-1 - 2x - 8}{x+4}$

$\qquad = \dfrac{-2x - 9}{x+4}$

Since $g(x)$ has the form $\dfrac{p(x)}{q(x)}$, where $p(x)$ and $q(x)$ are polynomials, $g(x)$ is a rational function.

The graph of the function $g(x) = \dfrac{-1}{x+4} - 2 = \dfrac{-1}{x-(-4)} + (-2)$ is a translation of the graph of $y = \dfrac{-1}{x}$ to the left 4 units ($h = -4$) and down 2 units ($k = -2$). The graph has a vertical asymptote at $x = -4$ and a horizontal asymptote at $y = -2$.

Tell whether each function is a rational function. Explain your answers.

1. $f(x) = \dfrac{x-3}{2x+5}$

2. $g(x) = \dfrac{2}{x+17}$

3. $h(x) = \dfrac{x^2 - 3}{x^2}$

4. $f(x) = \dfrac{3x}{\sqrt{5x}}$

5. $g(x) = \dfrac{3^x - 1}{3x - 1}$

6. $h(x) = \dfrac{-2}{x+3} - 10$

Describe the graph of each function.

7. $y = \dfrac{5}{x-3} + 2$

8. $y = \dfrac{-4}{x-1} - 5$

9. $y = \dfrac{-5}{x+7} - 2$

10. $y = \dfrac{6}{x+1.5} + 2.5$

You can use the standard equation $y = \dfrac{a}{x-h} + k$ to write the equation for a given graph that is a hyperbola with its asymptotes indicated. Since the asymptotes identify the values of h and k in the equation, you only need to know one point on the hyperbola so you can find the value of a.

Example 2

Find an equation of each hyperbola.

a.

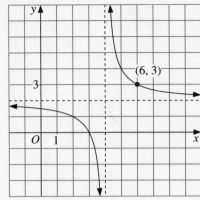

b.

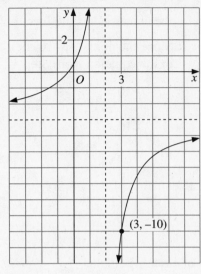

■ Solution ■

a. *Step 1*: Use the asymptotes to identify the values of h and k.

The asymptotes are $x = 4$ and $y = 2$, so $h = 4$ and $k = 2$. Therefore, the equation has the form $y = \dfrac{a}{x - 4} + 2$.

Step 2: Find the value of a.

Since the point $(6, 3)$ lies on the graph, substitute 6 for x and 3 for y in the equation from Step 1.

$$3 = \frac{a}{6 - 4} + 2$$

$$3 = \frac{a}{2} + 2$$

$$1 = \frac{a}{2}$$

$$a = 2$$

So, an equation of the graph is $y = \dfrac{2}{x - 4} + 2$.

(Solution continues on next page.)

b. *Step 1*: The asymptotes are $x = 2$ and $y = -3$, so $h = 2$ and $k = -3$. Therefore,

the equation has the form $y = \dfrac{a}{x-2} - 3$.

Step 2: Since the point $(3, -10)$ lies on the graph, substitute 3 for x and -10 for y in the equation from Step 1.

$$-10 = \frac{a}{3-2} - 3$$

$$-10 = \frac{a}{1} - 3$$

$$-7 = a$$

So, an equation of the graph is $y = \dfrac{-7}{x-2} - 3$.

Find an equation of each hyperbola.

11.

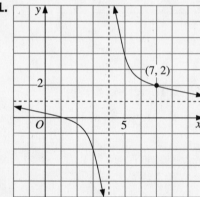

12.

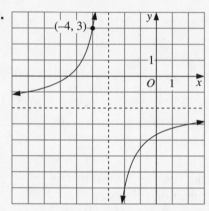

13.

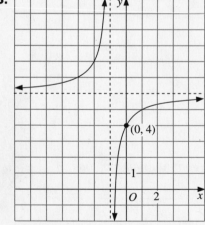

14.

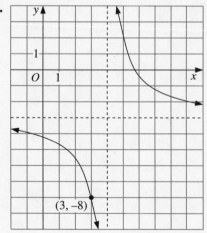

Rewriting rational functions

Given a rational function in the form $f(x) = \dfrac{p(x)}{q(x)}$, you can quickly indentify the asymptotes of its graph by rewriting it in the form $f(x) = \dfrac{a}{x-h} + k$.

Example 3

Rewrite the function $f(x) = \dfrac{-10x - 17}{x + 3}$ in the form $f(x) = \dfrac{a}{x-h} + k$. Then describe the graph of the function.

■ Solution ■

Step 1: Divide the denominator into the numerator.

$$
\begin{array}{r}
-10 \\
x + 3 \overline{)\,-10x - 17\,} \\
\underline{-10x - 30} \\
13
\end{array}
$$

So, $\dfrac{-10x - 17}{x + 3} = -10 + \dfrac{13}{x + 3}$.

Step 2: Write the function in the form $f(x) = \dfrac{a}{x-h} + k$.

$$f(x) = \frac{-10x - 17}{x + 3}$$

$$= -10 + \frac{13}{x + 3}$$

$$= \frac{13}{x + 3} - 10$$

$$= \frac{13}{x - (-3)} + (-10)$$

Since $h = -3$ and $k = -10$, there is a vertical asymptote at $x = -3$ and a horizontal asymptote at $y = -10$. The graph of $f(x)$ is the graph of $y = \dfrac{13}{x}$ translated 3 units to the left and 10 units down.

For each function:

a. Find the asymptotes of the function's graph.

b. Tell how the function's graph is related to a hyperbola with equation of the form $y = \dfrac{a}{x}$.

15. $f(x) = \dfrac{7x - 17}{x - 3}$ **16.** $f(x) = \dfrac{-5x - 7}{x + 1}$

17. $g(x) = \dfrac{-2x - 17}{x + 5}$ **18.** $g(x) = \dfrac{-8x + 26}{x - 2}$

TECHNOLOGY Refer to the rational function in the Application at the beginning of this section.

19. How long will it take the sound of thunder to travel 12,000 ft if the temperature is 65°F?

20. How does air temperature affect the speed of sound?

21. What distance will sound travel in 7 s if the air temperature is 32°F?

....
Spiral Review

For Exercises 22–24, tell whether *y* varies inversely with *x*. If so, state the constant of variation. *(Section 9.6)*

22. $y = \dfrac{\sqrt{10}}{x}$ **23.** $\dfrac{y}{x} = 1$ **24.** $\dfrac{xy}{5} = \dfrac{1}{10}$

25. A high school computer club recorded the following file sizes (in megabytes) for graphics files in their computer.

83.2, 85.8, 62.4, 85.8, 72.1, 77.9, 77.8, 81.7, 63.2, 75.7, 85.8

Find each statistic for these file sizes. *(Sections 6.4 and 6.5)*

a. range **b.** mode **c.** median **d.** mean

Divide. *(Section 9.2)*

26. $\dfrac{15x^2 - 11x - 12}{3x - 4}$ **27.** $\dfrac{2x^4 - 5x^3 + 12x^2 - 13x + 4}{2x^2 - 3x + 1}$

Working with General Rational Functions

Application

High levels of acidity in your mouth correspond to decreased pH levels, which can lead to cavities and tooth decay. The pH level in your mouth is related to the amount of time that has passed since you have eaten, and that relationship is a rational function.

UNDERSTANDING THE MAIN IDEAS

Vertical asymptotes and zeros of the denominator

When the numerator and denominator of a rational function have no common factors, the graph of the function has a vertical asymptote at each real solution of the equation formed by setting the denominator equal to 0.

Example 1

Find the vertical asymptotes of the graph of each rational function. Then match the function with one of these two graphs.

I.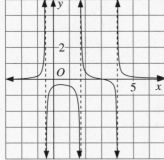

II.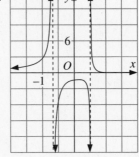

a. $f(x) = \dfrac{(x+3)(x-3)}{(x-4)(2x+1)(3x-5)}$

b. $g(x) = \dfrac{x^2 - 4x + 3}{6x^2 + x - 2}$

■ Solution ■

a. There are no common factors of the numerator and denominator, so the graph of f has a vertical asymptote at each real solution of the equation $(x - 4)(2x + 1)(3x - 5) = 0$. Using the zero product property,

$$x - 4 = 0 \qquad \text{or} \qquad 2x + 1 = 0 \qquad \text{or} \qquad 3x - 5 = 0$$

$$x = 4 \qquad\qquad\qquad x = -\frac{1}{2} \qquad\qquad\qquad x = \frac{5}{3}$$

Therefore, the graph has vertical asymptotes at $x = 4$, $x = -\frac{1}{2}$, and $x = \frac{5}{3}$. The graph of f is graph I.

b. First, write the numerator and denominator of g as a product of factors to see if they have any common factors.

$$g(x) = \frac{x^2 - 4x + 3}{6x^2 + x - 2} = \frac{(x - 3)(x - 1)}{(2x - 1)(3x + 2)}$$

Since there are no common factors of the numerator and denominator, the graph of g has a vertical asymptote at each real solution of the equation $(2x - 1)(3x + 2) = 0$. Using the zero product property,

$$2x - 1 = 0 \qquad \text{or} \qquad 3x + 2 = 0$$

$$x = \frac{1}{2} \qquad\qquad\qquad x = -\frac{2}{3}$$

Therefore, the graph has vertical asymptotes at $x = \frac{1}{2}$ and $x = -\frac{2}{3}$. The graph of g is graph II.

Find the vertical asymptotes of the graph of each function.

1. $f(x) = \dfrac{(x + 2)(x + 4)}{(x + 5)(x - 1)(x + 3)}$

2. $g(x) = \dfrac{x^2 - 25}{(x - 3)^2(x - 4)}$

3. $f(x) = \dfrac{x^2 - x - 2}{2x^2 - 5x - 3}$

4. $g(x) = \dfrac{4x}{x^2 - x - 2}$

Use what you know about vertical asymptotes to match each of the following graphs with a function from Exercises 1–4.

5.

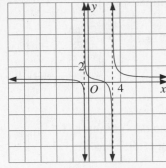

6.

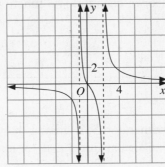

7.

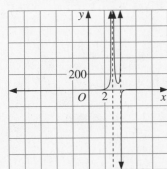

8.

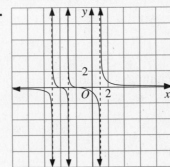

End behavior of a rational function

The end behavior of a rational function can be determined by comparing the degree of the numerator and the degree of the denominator. There are three possibilities for the end behavior of the function, depending on the relationship between these two degrees.

Let $f(x) = \dfrac{p(x)}{q(x)}$ where $p(x)$ is a polynomial of degree m and $q(x)$ is a polynomial of degree n. Let a be the leading coefficient of $p(x)$ and let b be the leading coefficient of $q(x)$.

- If $m < n$, then $f(x) \rightarrow 0$ as $x \rightarrow \pm\infty$. The graph of f has a horizontal asymptote at $y = 0$.

- If $m = n$, then $f(x) \rightarrow \dfrac{a}{b}$ as $x \rightarrow \pm\infty$. The graph of f has a horizontal asymptote at $y = \dfrac{a}{b}$.

- If $m > n$, then f has the same end behavior as the polynomial function $y = \dfrac{a}{b}x^{m-n}$. The graph of f has no horizontal asymptote.

In other words, this means that if the degree of the denominator of a rational function is greater than the degree of the numerator, then the function approaches (but does not reach) 0 as the value of x goes to positive and negative infinity; if the degrees are equal, then the function approaches the ratio of the leading coefficients as x goes to positive and negative infinity; and if the degree of the numerator is greater than the degree of the denominator, then you can use the techniques discussed in Section 9.3 to determine the end behavior.

Example 2

Describe the end behavior of each rational function using infinity notation.

a. $f(x) = \dfrac{3x^5 + 2x - 7}{-5x^3 + 5}$ **b.** $g(x) = \dfrac{7x^5 + 4x - 7}{-5x^5 - x^2 + 1}$ **c.** $h(x) = \dfrac{2x^3 + x^2 - 5}{4x^8 - 5x^4 + 2x - 3}$

■ Solution ■

a. The degree of the numerator, 5, is greater than the degree of the denominator, 3. Since the leading coefficients are 3 and –5, respectively, the function f has the same end behavior as the function $y = \dfrac{3}{-5}x^{5-3} = -\dfrac{3}{5}x^2$.

The end behavior is:

as $x \to +\infty, f(x) \to -\infty$; as $x \to -\infty, f(x) \to -\infty$.

b. The degrees of the numerator and denominator are the same. The leading coefficients are 7 and –5, respectively.

The end behavior is:

as $x \to \pm\infty, f(x) \to -\dfrac{7}{5}$.

c. The degree of the numerator, 3, is less than the degree of the denominator, 8.

The end behavior is:

as $x \to \pm\infty, f(x) \to 0$.

The following technique can also be used to determine the end behavior of a rational function. In general, a rational function has the same end behavior as the function formed by the quotient of the leading term of the numerator and the leading term of the denominator. When you simplify that quotient, you can examine the exponent and the sign of the coefficient to determine its end behavior as discussed in Section 9.3.

For each function, describe the function's end behavior using infinity notation.

9. $f(x) = \dfrac{3x + 1}{-5x^2 - 2x + 7}$ **10.** $f(x) = \dfrac{3x^5 - 2x^4 - 2}{-4x^5 - 3x}$

11. $g(x) = \dfrac{-5x^9 - 4}{(3x + 1)(2x + 5)}$ **12.** $g(x) = \dfrac{(x^5 - 1)(x^5 + 1)}{(x^3 - 7)(x^3 + 7)}$

For each function:

a. Find the vertical asymptotes of the function's graph.

b. Describe the function's end behavior using infinity notation.

c. Find any horizontal asymptotes of the function's graph.

13. $f(x) = \dfrac{-5x^4 - 7x}{2x^2 + x - 1}$ **14.** $g(x) = \dfrac{(x + 7)(3x - 9)}{4x^2 - 33x + 8}$

The Application at the beginning of this section refers to the rational function

$pH = \dfrac{65t^2 - 204t + 2340}{10t^2 + 360}$. Use this function for Exercises 15–17.

15. Without graphing, tell whether the graph of the function has any vertical asymptotes. Explain your answer.

16. What is the pH level when $t = 3$?

17. What is the approximate pH level when $t = 60$? when $t = 180$? as $t \rightarrow +\infty$?

....

Spiral Review

Tell how the graph of each function is related to a hyperbola with equation of the form $y = \dfrac{a}{x}$. *(Section 9.7)*

18. $f(x) = \dfrac{3}{x-5} + 9$

19. $f(x) = \dfrac{-2}{x+4} - 8$

20. $g(x) = \dfrac{2x+7}{x+5}$

21. $g(x) = \dfrac{-7x+74}{x-10}$

Solve each proportion. *(Toolbox, page 785)*

22. $\dfrac{3x}{8} = \dfrac{1.5}{4}$

23. $\dfrac{16}{5s-3} = \dfrac{4}{3}$

24. $\dfrac{7y-3}{4} = \dfrac{2y+1}{5}$

9.9

Solving Rational Equations

Learn how to . . .

- solve rational equations

So you can . . .

- solve problems involving rational equations

Application

When two people can complete a task at different rates, their combined rate is a rational function of their indivitual rates. You can find their combined rate by solving a rational equation.

Terms to Know

Example / Illustration

Rational equation (p. 450) an equation that contains only polynomials or quotients of polynomials	$\dfrac{7x+3}{4x-5}=2$ $\dfrac{x}{x-5}=4+\dfrac{2}{x-5}$ $\dfrac{2x-1}{x+7}+\dfrac{3}{3x+21}=5$

UNDERSTANDING THE MAIN IDEAS

The first step in solving a rational equation is to multiply each side of the equation by the least common denominator (LCD) of the rational expressions in the equation. You must check all possible solutions in the original equation because this procedure may introduce extraneous solutions. (*Note:* An extraneous solution will usually result in division by zero somewhere in the original equation.)

Example

Solve each rational equation.

a. $\dfrac{10x+21}{5x-3}=8$

b. $\dfrac{1}{m-3}+\dfrac{5}{m+2}=\dfrac{5}{m^3-m^2-6m}$

▪ Solution ▪

a. Multiply both sides of the equation by $5x - 3$.

$$\frac{10x + 21}{5x - 3}(5x - 3) = 8(5x - 3)$$

$$10x + 21 = 40x - 24$$

$$-30x = -45$$

$$x = 1.5$$

Check the solution in the original equation.

$$\frac{10(1.5) + 21}{5(1.5) - 3} \stackrel{?}{=} 8$$

$$\frac{15 + 21}{7.5 - 3} \stackrel{?}{=} 8$$

$$\frac{36}{4.5} \stackrel{?}{=} 8$$

$$8 = 8 \checkmark$$

The solution is 1.5.

b. First, factor the denominator on the right side and then find the LCD for the equation.

$$m^3 - m^2 - 6m = m(m^2 - m - 6)$$

$$= m(m - 3)(m + 2)$$

The LCD for the equation is $m(m - 3)(m + 2)$.

So multiply both sides of the equation by $m(m - 3)(m + 2)$.

$$\frac{1}{m - 3} + \frac{5}{m + 2} = \frac{5}{m(m - 3)(m + 2)}$$

$$\frac{m(m - 3)(m + 2)}{m - 3} + \frac{5m(m - 3)(m + 2)}{m + 2} = \frac{5m(m - 3)(m + 2)}{m(m - 3)(m + 2)}$$

$$m(m + 2) + 5m(m - 3) = 5$$

$$m^2 + 2m + 5m^2 - 15m = 5$$

$$6m^2 - 13m - 5 = 0$$

$$(2m - 5)(3m + 1) = 0$$

$$2m - 5 = 0 \quad \text{or} \quad 3m + 1 = 0$$

$$2m = 5 \quad \text{or} \quad 3m = -1$$

$$m = \frac{5}{2} \quad \text{or} \quad m = -\frac{1}{3}$$

Both possible solutions check in the original equation.

The solutions are $m = \frac{5}{2}$ and $m = -\frac{1}{3}$.

For Exercises 1–6, solve each equation.

1. $\dfrac{7x + 3}{4x - 5} = 2$

2. $\dfrac{6 - 3x}{8} = \dfrac{3x}{5}$

3. $\dfrac{x}{x - 5} = 4 + \dfrac{2}{x - 5}$

4. $\dfrac{5}{x} + \dfrac{3}{x - 1} = \dfrac{3}{x^2 - x}$

5. $\dfrac{2x - 1}{x + 7} + \dfrac{3}{3x + 21} = 5$

6. $\dfrac{3}{x + 4} + \dfrac{2x}{x - 1} = \dfrac{-12}{x^2 + 3x - 4}$

7. On many math exams you may see a problem like this: *A parking lot is covered with snow. One person with a snow blower can clear the lot in 3 h. One person with a snowplow can clear the lot in $\dfrac{3}{4}$ h. How long will it take to clear the lot if they work together?* Solve the problem.

8. **Mathematics Journal** Describe the relationship the number of tasks you can do in one minute and the number of minutes you need to do one complete task. Connect your response to the method you used to solve Exercise 7.

....
Spiral Review

Perform the indicated operation. *(Section 8.4)*

9. $(2 - 5i)(4 + i)$

10. $(5 - 6i)^2$

11. $(3 + 5i) - (14 - 7i)$

TECHNOLOGY Use a graphing calculator or software with matrix calculation capabilities to find an equation of the parabola passing through each set of points. *(Section 7.3)*

12. $(1, -2)$, $(2, -8)$, and $(3, -18)$

13. $(1, -4)$, $(2, -1)$, and $(3, 4)$

Chapter 9 Review ....

Complete these exercises for a review of Chapter 9. If you have difficulty with a particular problem, review the indicated section.

1. Use synthetic substitution to evaluate $5x^4 - 3x^3 - 32x^2 - 5x + 6$ for $x = 3$ and for $x = -2$. *(Section 9.1)*

2. Add $3x^3 + 2x^2 - 4x + 1$ and $6x^3 + 11x + 6$. *(Section 9.1)*

3. Subtract $6 - x - 6x^2 + x^3 + 5x^4$ from $3 + 2x - 5x^3 + 4x^4$. *(Section 9.1)*

For Exercises 4–7, find each product or quotient. *(Section 9.2)*

4. $(2x^2 - 4x + 1)(2x - 3)$

6. $(-4x^3 + 3x)(5x^2 - 3x + 7)$

5. $\dfrac{4x^3 - 4x^2 - 17x + 5}{2x - 5}$

7. $\dfrac{15x^3 + 5x^2 - 21x - 7}{3x + 1}$

8. Describe the end behavior of $f(x) = -2x^6 + 5x^3 - 2x + 3$. *(Section 9.3)*

9. Find all local maximums and minimums of $f(x) = -2x^6 + 5x^3 - 2x + 3$. *(Section 9.3)*

For Exercises 10 and 11, find all real and imaginary zeros of each function. Identify any double or triple zeros. *(Sections 9.4, 9.5)*

10. $f(x) = (x - 5)(x^2 + 7x + 10)$

11. $g(x) = (x^2 - 1)(x^2 - 3x + 5)$

12. List the possible rational zeros for the function $f(x) = 8x^3 + 3x^2 - 4x + 3$. *(Section 9.5)*

Tell whether the given values of x and y show inverse variation. If so, write an equation giving y as a function of x. *(Section 9.6)*

13.

x	y
-10	$\dfrac{1}{2}$
-5	1
5	-1
10	$-\dfrac{1}{2}$

14.

x	y
-10	-4
-5	-2
5	2
10	4

15.

x	y
-10	10
-5	5
5	-5
10	-10

16.

x	y
-10	$-\dfrac{1}{20}$
-5	$-\dfrac{1}{10}$
5	$\dfrac{1}{10}$
10	$\dfrac{1}{20}$

Find the asymptotes of the function's graph. Then describe the graph of each function in terms of the graph of $y = \dfrac{3}{x}$. *(Section 9.7)*

17. $y = \dfrac{3}{x + 8} + 7$

18. $y = \dfrac{-9x + 21}{x - 2}$

---

19. Let $g(x) = \dfrac{-2x^6 + 5x^3 - 2x + 3}{3x^2 - 7x + 4}$. Find the vertical asymptotes of the graph of g and describe the end behavior of g using infinity notation. *(Section 9.8)*

Solve and check. *(Section 9.9)*

20. $\dfrac{3x}{5x - 2} = \dfrac{3}{2x}$

21. $\dfrac{1}{x} + \dfrac{1}{x+1} = \dfrac{3}{x^2 + x}$

SPIRAL REVIEW   Chapters 1–9

For Exercises 1–3, solve for x.

1. $14x^2 - 33x - 5 = 0$

2. $\dfrac{2x + 3}{5x} = \dfrac{3x + 1}{7}$

3. $\begin{bmatrix} 3 & 1 \\ 5 & 2 \end{bmatrix}\begin{bmatrix} 2 & 8 \\ 3 & 9 \end{bmatrix} = \begin{bmatrix} x & 33 \\ 16 & 58 \end{bmatrix}$

4. Use synthetic substitution to evaluate $f(x) = -2x^4 + 3x^3 - 4x^2 + 5x - 1$ for $x = -2$.

Evaluate each expression.

5. $\left(\dfrac{1}{25}\right)^{3/2}$

6. $\log_8 4$

7. $\dfrac{15^0}{36^{3/2}}$

Find the zeros of each function.

8. $f(x) = (x^2 - 4)(x^2 - 5x + 6)$

9. $g(x) = x^4 - 81$

Write an equation in slope-intercept form for the line passing through each pair of points.

10. $(2, -9)$ and $(-3, 1)$

11. $(4, -12)$ and $(-2, 6)$

Rewrite each equation in vertex form, $y = a(x - h)^2 + k$.

12. $y = 3x^2 - 30x + 77$

13. $y = -4x^2 - 16x - 24$

Rewrite each equation in the form $y = \dfrac{a}{x - h} + k$.

14. $y = \dfrac{-2x + 13}{x - 5}$

15. $y = \dfrac{7x + 5}{x + 1}$

Graph each system of equations or inequalities.

16. $y = x + 6$
   $y = -2x - 3$

17. $y \geq 2x$
   $x > -3$
   $y < -x$

Section

10.1

GOAL

Learn how to . . .

- find and use formulas for sequences

So you can . . .

- make predictions

Sequences

Application

The spider follows a pattern as it builds its web. The lengths of each strand of the web form a sequence.

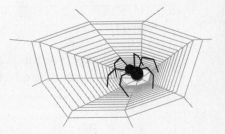

Terms to Know	**Example / Illustration**
Sequence (p. 465) a set of ordered elements	1, 3, 5, 7, 9
Term (p. 465) one of the elements of a sequence	The sequence 1, 3, 5, 7, 9 has five terms.
Finite sequence (p. 467) a sequence that has a last term	The sequence 1, 3, 5, 7, 9 is a finite sequence.
Infinite sequence (p. 467) a sequence that continues without end	The odd numbers 1, 3, 5, 7, 9, ... form an infinite sequence.

UNDERSTANDING THE MAIN IDEAS

When a number sequence has a pattern, you may be able to write a rule for the relationship between the terms of the sequence and their position in the sequence. For such a sequence, the rule defines each term t as a function of its position n. The rule $t_n = 3n$ represents a sequence in which each term is 3 times its position value. (*Note*: t_n is read "t sub n.")

Example 1

Use the sequence 120, 60, 40, 30, ...

a. Find the next three terms of the sequence.

b. Write a formula for the nth term, t_n, of the sequence in terms of n.

■ Solution ■

a. Do not expect to figure out the pattern right away. Try a variety of ideas, such as looking at sums, products, quotients, powers, and so on.

Position in sequence	1	2	3	4	...
Term	$120 = \dfrac{120}{1}$	$60 = \dfrac{120}{2}$	$40 = \dfrac{120}{3}$	$30 = \dfrac{120}{4}$	...

Each term is the quotient of 120 divided by the position of the term. The next three terms are $\dfrac{120}{5} = 24$, $\dfrac{120}{6} = 20$, and $\dfrac{120}{7} = 17\dfrac{1}{7}$.

b. Since each term is the quotient of 120 and the position value of the term, the formula is $t_n = \dfrac{120}{n}$.

For Exercises 1–5, find the next 3 terms of each sequence, if possible. If not, explain why.

1. 0.1, 0.01, 0.001, 0.0001, ...

2. 7, 171, 71717, 1717171, ...

3. –2, 4, –8, 16, ...

4. leap years: 1996, 2000, 2004, 2008, ...

5. the sequence of the last two digits of the telephone numbers in the first column of page 12 of your local telephone book

6. Open-ended Write three different sequences that begin 4, 2, ... Describe the pattern in each sequence.

Using the formula for a sequence

For any sequence formula, such as $t_n = 3n$, the domain consists of the counting numbers 1, 2, 3, ..., while the range consists of the terms of the sequence.

Example 2

Find the first 4 terms and the 10th term of the sequence $t_n = (-1)^n(2n)$.

Study Guide, ALGEBRA 2: EXPLORATIONS AND APPLICATIONS

■ Solution ■

Substitute 1, 2, 3, 4, and 10 for n in the formula for t_n.

$$t_1 = (-1)^1(2(1)) = -2 \qquad t_2 = (-1)^2(2(2)) = 4 \qquad t_3 = (-1)^3(2(3)) = -6$$

$$t_4 = (-1)^4(2(4)) = 8 \qquad t_{10} = (-1)^{10}(2(10)) = 20$$

(Notice how the factor $(-1)^n$ makes the signs of the first four terms of the sequence alternate between positive and negative.)

The first 4 terms of the sequence are –2, 4, –6, and 8. The 10th term is 20.

---

7. Suppose the sequence formula in Example 2 is changed to $t_n = (-1)^{n+1}(2n)$. How do the first 4 terms and the 10th term of the sequence change from those found in Example 2?

Find the first 4 terms and the 10th term of each sequence.

8. $t_n = n^2 - n$ **9.** $t_n = \dfrac{n}{n+1}$ **10.** $t_n = 5^{n-2}$ **11.** $t_n = \log 10^n$

Write a formula for t_n.

12. $1, \dfrac{1}{2}, \dfrac{1}{3}, \dfrac{1}{4}, \ldots$

13. 2, –2, 2, –2, …

14. 1, 4, 9, 25, …

15. $\sqrt{3}, 2, \sqrt{5}, \sqrt{6}, \ldots$

Tell whether each sequence is *finite* or *infinite*.

16. the positive even numbers less than 1000

17. the dates of the Saturdays in March

18. the common logarithms of the counting numbers

. . . .
Spiral Review

Solve each rational equation. *(Section 9.9)*

19. $\dfrac{1}{x} = \dfrac{5}{10x-4}$ **20.** $\dfrac{y+1}{5} = \dfrac{36}{y-10}$ **21.** $\dfrac{4t-8}{3t} + \dfrac{24}{t} = 4$ **22.** $\dfrac{1}{x+5} = \dfrac{1}{x^2-25}$

Simplify each radical expression. *(Section 8.2)*

23. $\sqrt{50x^{12}}$ **24.** $\sqrt[3]{625}$ **25.** $\sqrt[4]{80x^{10}}$ **26.** $\sqrt[3]{\dfrac{1}{8}y^{12}}$

Sketch the graph of each equation. *(Sections 2.2 and 3.3)*

27. $y = -\dfrac{3}{2}x + 6$ **28.** $y = \dfrac{1}{10}(4^x)$ **29.** $y = 4x - 20$ **30.** $y = (0.75)^x$

Section 10.2

Arithmetic and Geometric Sequences

Learn how to . . .

- determine whether a sequence is arithmetic or geometric

- find an arithmetic or geometric mean

So you can . . .

- find specific terms of a sequence

Application

Each time the ball bounces, it rebounds to $\frac{3}{4}$ of its previous height. The sequence of heights is a geometric sequence with common ratio $\frac{3}{4}$.

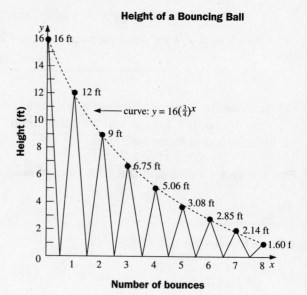

Height of a Bouncing Ball

curve: $y = 16(\frac{3}{4})^x$

16 ft, 12 ft, 9 ft, 6.75 ft, 5.06 ft, 3.08 ft, 2.85 ft, 2.14 ft, 1.60 f

Height (ft) / Number of bounces

Terms to Know

Example / Illustration

Geometric sequence (p. 473)
a sequence in which the ratio of any term to the term before it is constant

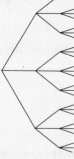

The number of branches at each step of this pattern form the geometric sequence 3, 9, 27.

Common ratio (p. 473)
the constant ratio in a geometric sequence

The common ratio of the geometric sequence represented above is 3.

Arithmetic sequence (p. 473) a sequence in which the difference between any term and the term before it is constant	The dates of the Mondays during the month below form an arithmetic sequence.

S	M	T	W	Th	F	S
	①	2	3	4	5	6
7	⑧	9	10	11	12	13
14	⑮	16	17	18	19	20
21	㉒	23	24	25	26	27
28	㉙	30	31			

Common difference (p. 473) the constant difference in an arithmetic sequence	The common difference of the arithmetic sequence represented above is 7.
Geometric mean (p. 475) for any two positive terms a and b, the positive square root of the product ab; $x = \sqrt{ab}$ (The sequence a, x, b is a geometric sequence.)	The geometric mean of 3 and 27 is $\sqrt{3 \cdot 27} = \sqrt{81} = 9$.
Arithmetic mean (p. 475) for any two terms a and b, the *average* or *mean* of a and b; $x = \dfrac{a+b}{2}$ (The sequence a, x, b is an arithmetic sequence.)	The arithmetic mean of 3 and 27 is $\dfrac{3+27}{2} = \dfrac{30}{2} = 15$.

UNDERSTANDING THE MAIN IDEAS

Arithmetic sequences and their formula

The terms of an arithmetic sequence increase (or decrease) by a constant amount, d. The general formula for an arithmetic sequence is $t_n = t_1 + d(n - 1)$, where t_1 is the first term, d is the common difference, n is the position of the term in the sequence, and t_n is the nth term of the sequence. If the first term and the common difference of an arithmetic sequence are known (or can be determined), then a formula expressing t as a function of n can be written.

Example 1

Use the sequence 6, 11, 16, 21, ...

a. Show this sequence is arithmetic.   **b.** Write a formula for t_n.

■ Solution ■

a. To determine if a sequence is arithmetic, check the difference between each term and the previous term. For the sequence to be arithmetic, these differences must all be the same.

$$6, \qquad 11, \qquad 16, \qquad 21, \quad ...$$
$$11 - 6 = 5 \quad 16 - 11 = 5 \quad 21 - 16 = 5$$

Since each difference is 5, the sequence is an arithmetic sequence with a common difference of 5.

b. Since $t_1 = 6$ and $d = 5$, the formula is $t_n = 6 + 5(n - 1)$, or $t_n = 5n + 1$.

---

1. Use $n = 1, 2, 3$, and 4 to verify that the formula found in part (b) of Example 1 is correct.

2. Find the 21st term of the sequence given in Example 1.

For Exercises 3–5, verify that each sequence is arithmetic. Give the common difference and a formula for t_n.

3. 100, 96, 92, 88, ...   4. –15, –10, –5, 0, ...   5. 2.03, 2.24, 2.45, 2.66, ...

6. Find the 12th term of each of the sequences given in Exercises 3–5.

Geometric sequences and their formula

Successive terms of a geometric sequence are in a constant ratio, r. That is, the ratio of any term to the term preceding it is constant. The general formula for a geometric sequence is $t_n = t_1 \cdot r^{n-1}$, where t_1 is the first term, r is the common ratio, n is the position of the term in the sequence, and t_n is the nth term of the sequence. If the first term and the common ratio of a geometric sequence are known (or can be determined), then a formula expressing t as a function of n can be written.

Example 2

Use the sequence 80, 40, 20, 10, ...

a. Show that this sequence is geometric.   **b.** Write a formula for t_n.

■ Solution ■

a. To determine if a sequence is geometric, check the ratio of each term to the previous term. For the sequence to be geometric, these ratios must all be the same.

80,   40,   20,   10,   ...

$\frac{40}{80} = 0.5$ $\frac{20}{40} = 0.5$ $\frac{10}{20} = 0.5$

The sequence is a geometric sequence with a common ratio of 0.5.

b. Since $t_1 = 80$ and $r = 0.5$, the formula is $t_n = 80(0.5)^{n-1}$.

---

7. Use $n = 1, 2, 3,$ and 4 to verify that the solution found in part (b) of Example 2 is correct.

8. Find the 8th term of the sequence given in Example 2.

For Exercises 9–11, verify that each sequence is geometric. Give the common ratio and a formula for t_n.

9. 6, 9, 13.5, 20.25, ...

10. $5, 5\sqrt{5}, 25, 25\sqrt{5}, ...$

11. –1, 3, –9, 27, ...

12. Find the 8th term of each of the sequences given in Exercises 9–11.

13. The first two terms of a sequence are 12 and –4. Find the third and fourth terms, assuming that the sequence is (**a**) arithmetic and (**b**) geometric.

Tell whether each sequence is *arithmetic, geometric,* or *neither.*

14. $\frac{1}{2}, \frac{1}{10}, \frac{1}{50}, \frac{1}{250}, ...$

15. $\frac{1}{2}, \frac{2}{3}, \frac{3}{4}, \frac{4}{5}, ...$

16. $1 + i, 2, 3 - i, 4 - 2i, ...$

For each pair of numbers in Exercises 17–20:

a. Find the arithmetic mean.

b. Find the geometric mean.

17. 4, 25

18. $\frac{2}{5}, \frac{5}{9}$

19. 15, 16.2

20. 1.01, 4.04

21. Mathematics Journal In what ways are arithmetic and geometric sequences alike? In what ways are they different?

....

Spiral Review

Find the 11th term of each sequence. *(Section 10.1)*

22. $i, 2i, 3i, 4i, ...$

23. $1, \frac{1}{2}, \frac{1}{3}, \frac{1}{4}, ...$

24. 7.7, 7.77, 7.777, 7.7777, ...

Find the zeros of each function. *(Section 9.4)*

25. $f(x) = 3(x - 2)(x + 8)(x - 1)$

26. $f(x) = x^3 - 2x^2 - 6x + 12$

Exploring Recursion

Learn how to ...

▪ find the terms of a
sequence defined
recursively

▪ write a recursive
formula for a
sequence

So you can ...

▪ identify sequences
recursively

Application

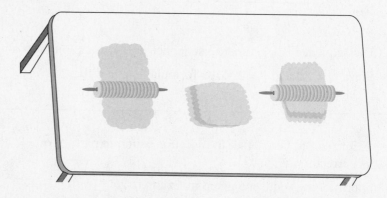

A baker uses recursion when making flaky puff pastry. He rolls out the dough,
butters it, folds it in thirds, and then turns the pastry 90°. The process is then
repeated several times.

Terms to Know

Example / Illustration

Explicit formula (p. 480) a formula for a sequence in which the nth term, t_n, is expressed as a function of its position number n	The sequence of odd numbers are given by the explicit formula $t_n = 1 + 2(n - 1)$.

n	t_n
1	$1 + 2(1 - 1) = 1$
2	$1 + 2(2 - 1) = 3$
3	$1 + 2(3 - 1) = 5$
4	$1 + 2(4 - 1) = 7$
...	...

Recursion (p. 481) a process of generating the terms of a sequence by beginning with an initial term and then finding additional terms by repeatedly applying the same operation to the previous term	You can generate the odd numbers by recursion: each odd number equals the previous odd number plus 2.

Terms to Know	Example / Illustration

Recursive formula (p. 481)
a formula with two parts: a starting value, and an equation for the nth term, t_n, as a function of t_{n-1}, the preceding term (Some recursive formulas give the first *two* terms and an equation that relates the nth term to the *two* preceding terms.)

A recursive formula for the sequence of odd numbers is $t_1 = 1, t_n = t_{n-1} + 2$.

n	t_n
1	1
2	$1 + 2 = 3$
3	$3 + 2 = 5$
4	$5 + 2 = 7$
...	...

UNDERSTANDING THE MAIN IDEAS

Finding the terms of a sequence using a recursive formula

Since a recursive formula assigns a starting value, the first term of the sequence is given. The recursion equation given as the second part of the formula can then be used to find the second term by substituting the value of the first term into the equation. The third term can then be found by substituting the value of the second term into the equation, and so on.

Example 1

Find the first 4 terms of the sequence $t_1 = -2, t_n = (t_{n-1})^2 - 5$.

■ Solution ■

Each time you find a term of the sequence, use it to find the next term.

n	t_n
1	-2
2	$(-2)^2 - 5 = -1$
3	$(-1)^2 - 5 = -4$
4	$(-4)^2 - 5 = 11$
...	...

The first four terms are -2, -1, -4, and 11.

Find the first 4 terms of each sequence.

1. $t_1 = 8$; $t_n = -1.5(t_{n-1})$

2. $t_1 = 16$; $t_n = \sqrt{t_{n-1}}$

Writing a recursive formula for a sequence

Many types of sequences have recursive formulas. Here are the recursive formulas for arithmetic and geometric sequences.

Arithmetic: t_1 = starting value; $t_n = t_{n-1} + d$

Geometric: t_1 = starting value; $t_n = r(t_{n-1})$

Recall that the explicit formula for an arithmetic sequence is $t_n = t_1 + d(n-1)$
and that the explicit formula for a geometric sequence is $t_n = t_1(r^{n-1})$.

Example 2

Write a recursive formula for t_n, the number of layers of pastry, at steps n of the process for making puff pastry described in the Application.

■ Solution ■

Step 1: Find the value of the first term, t_1.

You start with one layer of dough, so $t_1 = 1$.

Step 2: Look for a relationship between each term and the next term. Use this relationship to write the recursion equation.

The pastry is folded into thirds during the first stage of the process, so there are then 3 layers of dough. Thus, $t_2 = 3$.

When the pastry is folded into thirds during the second stage of the process, there are 3×3, or 9 layers of dough. Thus, $t_3 = 9$.

When the pastry is folded into thirds during the third stage of the process, there are 3×9, or 27 layers of dough. Thus, $t_4 = 27$.

So the number of layers of dough after each stage of the process is three times the number of layers at the previous stage. Therefore, the recursion equation is

$$t_n = 3(t_{n-1}).$$

The recursive formula is $t_1 = 1$, $t_n = 3(t_{n-1})$.

---

3. What type of sequence is the sequence found in Example 2? Write an explicit formula for the sequence.

4. Another way to make puff pastry is to repeatedly butter the dough, roll it out, fold it in half, and then turn it 90°. Write a recursive formula for t_n, the number of pastry layers.

Write a recursive formula for each sequence.

5. 6, 2, –2, –6, ...

6. 2.4, 24, 240, 2400, ...

7. 1, 2, 5, 26, ...

8. $t_n = 4n - 8$

9. $t_n = -5(-0.1)^{n-1}$

For each sequence:

a. Find the first 4 terms.

b. Write an explicit formula.

10. $t_1 = 9, t_n = \frac{2}{3}(t_{n-1})$
11. $t_1 = i, t_n = 1 - i + t_{n-1}$
12. $t_1 = \sqrt{7}, t_n = -\sqrt{7}(t_{n-1})$

Marcia Garth has an investment called an *annuity*. This is how the investment works. Every month her balance is increased by 0.75% because of interest. She then withdraws $400. Suppose her initial investment was $5000.

13. Write a recursive formula that will give t_n, the amount she has in her account, after the nth withdrawal.

14. Technology Use a graphing calculator or spreadsheet software to determine the number of months until there is no money left in the account.

....
Spiral Review

Find the geometric mean of each pair of numbers. *(Section 10.2)*

15. 8, 52
16. 5, 35
17. $\frac{2}{3}, \frac{8}{27}$
18. $6\sqrt{30}, 2\sqrt{30}$

Tell whether the data that can be gathered about each variable are *categorical* or *numerical*. Then describe the categories or numbers. *(Section 6.1)*

19. adult glove sizes
20. passenger cars
21. newspaper ads

Evaluate each expression when $a = 3$ and $b = -4$. *(Toolbox, page 780)*

22. $3(a + 7b)$
23. $\frac{b(a-5)}{12}$
24. $\frac{-6(2a+b)}{ab}$
25. $3a - 5(b + 8)$

Sums of Arithmetic Series

Learn how to . . .

- use a formula to find the sum of an arithmetic series
- use sigma notation for series

So you can . . .

- explore finite arithmetic series

Application

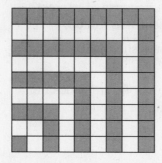

The total number of dark tiles in this pattern is given by the arithmetic series $1 + 5 + 9 + 13 + 17$.

Terms to Know	Example / Illustration
Series (p. 488) the sum of the terms of a sequence	$1 + 4 + 9 + 16 + 25 + 36$
Arithmetic series (p. 488) the indicated sum of an arithmetic sequence	$1 + 5 + 9 + 13 + 17$

UNDERSTANDING THE MAIN IDEAS

Sum of a finite arithmetic series

The general formula for the sum S_n of the first n terms of a finite arithmetic series $t_1 + t_2 + t_3 + \cdots + t_n$ is:

$$S_n = \frac{n}{2}(t_1 + t_n)$$

In words, the sum of the first n terms of a finite arithmetic sequence is the product of half the number of terms and the sum of the first term and the nth term.

Example 1

Find each sum.

a. $-2 + (-6) + (-10) + (-14) + (-18) + (-22)$

b. $8 + 11 + 14 + 17 + \cdots + 332$

■ Solution ■

a. This series is based on the sequence $-2, -6, -10, -14, -18, -22$, which is an arithmetic sequence with common difference -4. The series has six terms.

The sum is $S_6 = \frac{6}{2}[(-2) + (-22)] = 3(-24) = -72$.

b. This problem is harder than the one in part (a) because you must determine the number of terms in the series before using the formula for the sum.

The series is based on the sequence $8, 11, 14, 17, \ldots, 332$ which is an arithmetic sequence with common difference 3. So, you can use the explicit formula for an arithmetic sequence to find the value of n. We know that $t_1 = 8$, $d = 3$, and $t_n = 332$.

$$t_n = t_1 + d(n - 1)$$
$$332 = 8 + 3(n - 1)$$
$$332 = 8 + 3n - 3$$
$$332 = 3n + 5$$
$$327 = 3n$$
$$109 = n$$

Now you can use the general formula for the sum of an infinite arithmetic series, with $n = 109$, $t_1 = 8$, and $t_n = 332$.

The sum is $S_{109} = \frac{109}{2}(8 + 332) = \frac{109}{2}(340) = 18{,}530$.

Find the sum of each series.

1. $10 + 20 + 30 + 40 + \cdots + 120$

2. $-25 + (-5) + 15 + 35 + 55 + 75 + 95$

3. $-25 + (-5) + 15 + \cdots + 895$

4. $88 + 82 + 76 + 70 + \cdots + 4$

The first four terms and the sum of a finite arithmetic series are given. Find the last term of each series.

5. $2 + 2.25 + 2.5 + 2.75 + \cdots + \underline{\ ?\ }$; $S_{29} = 159.5$

6. $1 + 6 + 11 + 16 + \cdots + \underline{\ ?\ }$; $S_{32} = 2512$

7. $18 + 9 + 0 + (-9) + \cdots + \underline{\ ?\ }$; $S_{11} = -297$

To improve his fitness, Benoit plans to run regularly. The first week, he plans to run a total of 4 mi. The next week, he plans to run 4.5 mi. Every week his total mileage will be 0.5 mi greater than his mileage the week before. Use this information for Exercises 8 and 9.

8. Write Benoit's total mileage during the first 13 weeks of his fitness program as a finite arithmetic series.

9. How many total miles will he run during the first 13 weeks? during weeks 14 through 26?

---

10. Writing Zoe said "The sum of a finite arithmetic series is the number of terms times the average of the first and last terms." Is Zoe correct? Explain your answer.

Using sigma notation

A series can be written in a compact form, called *sigma notation*, by using the summation symbol $\sum$ (the Greek letter "sigma"). For example, the expression $\sum_{n=1}^{5} 4n$ represents the sum of the values of $4n$ for integer values of n from 1 to 5. To write this series in *expanded form*, you substitute each of the integers 1 through 5 for n into the formula $4n$ and show the series as the sum of the five results.

$$\sum_{n=1}^{5} 4n = 4(1) + 4(2) + 4(3) + 4(4) + 4(5)$$
$$= 4 + 8 + 12 + 16 + 20$$

Example 2

Expand the series $\sum_{n=2}^{6} (n^2 - 1)$.

■ Solution ■

The numbers above and below the summation symbol indicate that the first value of n is 2 and the last value is 6, so substitute the integers 2, 3, 4, 5, and 6 for n in the formula. Then write the results as a series.

$$\sum_{n=2}^{6} (n^2 - 1) = (2^2 - 1) + (3^2 - 1) + (4^2 - 1) + (5^2 - 1) + (6^2 - 1)$$
$$= (4 - 1) + (9 - 1) + (16 - 1) + (25 - 1) + (36 - 1)$$
$$= 3 + 8 + 15 + 24 + 35$$

The expanded form of the series is $3 + 8 + 15 + 24 + 35$.

11. Can you use the formula for the sum of a finite arithmetic series to find the sum of the series in Example 2? Explain.

For Exercises 12–15, expand each series.

12. $\sum_{n=1}^{6} 0.1n^3$ **13.** $\sum_{n=1}^{8} (3n - 10)$ **14.** $\sum_{j=1}^{4} (-2j^2 + 3)$ **15.** $\sum_{k=1}^{5} [80 - 4(k - 1)]$

16. Which of the series in Exercises 12–15 are arithmetic? Find the sum of each arithmetic series.

Write each series in sigma notation and find the sum.

17. $2 + 5 + 8 + 11 + \cdots + 299$

18. $1 + 1.2 + 1.4 + 1.6 + \cdots + 2.8$

19. $(-1) + (-11) + (-21) + (-31) + \cdots + (-201)$

....
Spiral Review

Write a recursive formula for each sequence. *(Section 10.3)*

20. $4, -2, 1, -\dfrac{1}{2}, \ldots$

21. $1, 9, 17, 25, \ldots$

22. $625, 25, 5, \sqrt{5}, \ldots$

Multiply. *(Section 1.4)*

23. $\begin{bmatrix} 6 & 1 & -2 & 7 \end{bmatrix} \begin{bmatrix} 3 \\ -4 \\ 0 \\ 8 \end{bmatrix}$

24. $\begin{bmatrix} 0 & 1 \\ 3 & -2 \end{bmatrix} \begin{bmatrix} 3 \\ 4 \end{bmatrix}$

25. $\begin{bmatrix} 5 & 2 & -1 & 0 \\ 6 & -3 & 0 & 2 \\ -1 & 0 & 8 & -5 \end{bmatrix} \begin{bmatrix} 3 & 1 \\ 2 & -5 \\ 6 & 0 \\ -1 & -4 \end{bmatrix}$

Sums of Geometric Series

Learn how to . . .

- use a formula to find the sum of a finite geometric series

- use a formula to find the sum of an infinite geometric series, if the sum exists

So you can . . .

- explore the uses of geometric series

Application

When written in expanded form, a repeating decimal can be seen as an infinte geometric series that has a sum.

$$0.44444\ldots = 0.4 + 0.04 + 0.004 + 0.0004 + 0.00004 + \cdots = \frac{4}{9}$$

Terms to Know	**Example / Illustration**
Geometric series (p. 495) the indicated sum of a geometric sequence	$0.44444\ldots = \dfrac{4}{9} = \displaystyle\sum_{n=1}^{\infty} 0.4(10^{n-1})$

UNDERSTANDING THE MAIN IDEAS

Finding the sum of a finite geometric series

The general formula for the sum S_n of the finite geometric series
$t_1 + t_1 \cdot r + t_1 \cdot r^2 + t_1 \cdot r^3 + \cdots + t_1 \cdot r^{n-1}$ where $r \neq 1$ is

$$S_n = \frac{t_1(1 - r^n)}{1 - r}.$$

Example 1

When Andrea Donnelly graduates from college, she has agreed to make annual donations to the alumni fund. The size of the donations will increase each year. The first year, she will donate $50. Every year after that, her donation will be 15% larger than her donation the previous year.

a. Use sigma notation to write the total amount Andrea donates in n years.

b. Approximately how much money will Andrea donate in the first 25 years after her graduation?

▪ Solution ▪

a. Find the pattern in Andrea's donations. Recall that a 15% increase corresponds to a growth factor of $100\% + 15\% = 115\%$, or 1.15.

Year	1	2	3	4	⋯
Donation	$50	$50(1.15) =$ $57.50	$57.50(1.15)$ $\approx \$66.13$	$66.125(1.15)$ $\approx \$76.04$	⋯

The donations form a geometric sequence. The common ratio is 1.15. The sum of the first n donations is $50 + 50(1.15) + 50(1.15)^2 + 50(1.15)^3 + \cdots + 50(1.15)^{n-1}$, which in sigma notation is

$$\sum_{k=1}^{n} 50(1.15)^{k-1}.$$

b. Use the formula for the sum of a finite geometric series, with $n = 25$, $t_1 = 50$, and $r = 1.15$.

$$S_n = \frac{t_1(1 - r^n)}{1 - r}$$

$$S_{25} = \frac{50(1 - 1.15^{25})}{1 - 1.15}$$

$$\approx \frac{50(1 - 32.9)}{-0.15} \quad \leftarrow \text{Use a calculator.}$$

$$\approx \frac{50(-31.9)}{-0.15}$$

$$\approx 10{,}600$$

Over the 25 years, Andrea will donate approximately $10,600.

---

1. Andrea's college roomate Cassandra has chosen a different donation plan. She will donate $150 the first year and increase her donation by 10% each year. How will the total of her donations during the first 25 years after graduation compare to Andrea's total?

2. Which woman will donate the most money during the first 35 years after her graduation?

Find the sum of each series. (*Hint*: For Exercises 3 and 4, use the explicit formula for a geometric sequence to find the value of *n* before attempting to find the sum.)

3. $2 + 6 + 18 + \cdots + 4374$

4. $i + 2i + 4i + 8i + 16i + \cdots + 1024i$

5. $8 + 8(1.06) + 8(1.06)^2 + 8(1.06)^3 + \cdots + 8(1.06)^{12}$

6. $\displaystyle\sum_{n=1}^{9} 5(0.4)^{n-1}$

7. $\displaystyle\sum_{n=1}^{5} 32\left(\frac{7}{8}\right)^{n-1}$

8. $\displaystyle\sum_{n=1}^{8} -6(-4)^{n-1}$

Finding the sum of an infinite geometric series

The general form for the sum of an infinite geometric series
$t_1 + t_1 \cdot r + t_1 \cdot r^2 + t_1 \cdot r^3 + \cdots$ where $|r| < 1$ is

$$S = \frac{t_1}{1 - r}.$$

If $|r| \geq 1$, the infinite geometric series does not have a sum.

Example 2

Find the sum of each infinite geometric series, if the sum exists.

a. $2 + (-1) + \frac{1}{2} + \left(-\frac{1}{4}\right) + \frac{1}{8} + \cdots$ **b.** $\sum_{n=1}^{\infty} 50(1.15)^{n-1}$

■ Solution ■

a. The common ratio of this infinite geometric series is $-\frac{1}{2}$. Since $\left|-\frac{1}{2}\right| < 1$, the series has a sum.

$$S = \frac{t_1}{1 - r} \qquad \leftarrow t_1 = 2 \text{ and } r = -\frac{1}{2}$$

$$= \frac{2}{1 - \left(-\frac{1}{2}\right)}$$

$$= \frac{2}{\frac{3}{2}}$$

$$= \frac{4}{3}$$

The sum of the series is $\frac{4}{3}$.

b. From the sigma notation, it can be seen that the common ratio is 1.15. Since $|1.15| \geq 1$, this series does not have a sum.

Find the sum of each infinite geometric series, if the sum exists. If a series does not have a sum, explain why not.

9. $30 + 24 + 19.2 + 15.36 + \cdots$ **10.** $(-24) + 18 + (-13.5) + 10.125 + \cdots$

11. $\frac{1}{16} + \frac{1}{4} + 1 + 4 + \cdots$ **12.** $(-6) + 6 + (-6) + 6 + \cdots$

13. $\sum_{n=1}^{\infty} 5(\sqrt{5})^n$ **14.** $\sum_{n=1}^{\infty} 1000(0.98)^n$ **15.** $\sum_{n=1}^{\infty} \frac{3}{4}\left(\frac{8}{9}\right)^{n-1}$

Study Guide, ALGEBRA 2: EXPLORATIONS AND APPLICATIONS

For Exercises 16–18, express each decimal as a fraction. (*Hint*: First write the repeating decimal as an infinite geometric series and then find the sum of the series.)

16. $0.0666\ldots$ **17.** $-0.534534534\ldots$ **18.** $1.41414\ldots$

19. Mathematics Journal Explain why an infinite geometric series cannot have a sum if $r > 1$.

················

Spiral Review

Write each series in sigma notation and find the sum. *(Section 10.4)*

20. $4.2 + 4.7 + 5.2 + 5.7 + \cdots + 7.2$ **21.** $55 + 49 + 43 + 37 + \cdots + (-53)$

Solve each system of equations. *(Section 7.2)*

22. $5x - 3y = 9$
$x + y = 5$

23. $4x + 7y = -1$
$2x - y = 13$

24. $3x + y = -18$
$x + 3y = 10$

25. $y = 2x + 20$
$y = -3x + 5$

Find the slope of the line passing through each pair of points. *(Section 2.2)*

26. $(0, 0)$ and $(-3, 7)$ **27.** $(5, 9)$ and $(0, 3)$

28. $(2, -8)$ and $(4, -8)$ **29.** $(6, -6)$ and $(-3, -5)$

Chapter 10 Review ....

Complete these exercises for a review of Chapter 10. If you have difficulty with a particular problem, review the indicated section.

Find the first 5 terms and the 12th term of each sequence. Round your answers to the nearest hundredth. *(Section 10.1)*

1. $t_n = 2n^2 - 5$

2. $t_n = \ln 10n$

Write a formula for t_n. *(Section 10.1)*

3. $\dfrac{6}{5}, \dfrac{7}{6}, \dfrac{8}{7}, \dfrac{9}{8}, \ldots$

4. $\pi, 4\pi, 9\pi, 16\pi, \ldots$

For Exercises 5–7, tell whether the sequence is *arithmetic*, *geometric*, or *neither*. If the sequence is arithmetic or geometric, give a formula for t_n. *(Section 10.2)*

5. $\dfrac{1}{81}, \dfrac{1}{27}, \dfrac{1}{9}, \dfrac{1}{3}, \ldots$

6. 1.2, 1.23, 1.234, ...

7. 0.16, 0.20, 0.24, 0.28, ...

8. Open-ended Give two pairs of numbers whose arithmetic mean is 8. Then give two other pairs of numbers whose geometric mean is 8. *(Section 10.2)*

Toothpicks were used to create the triangular designs shown below. Use the figures for Exercises 9 and 10. *(Section 10.3)*

Stage 1 Stage 2 Stage 3 Stage 4

9. How many toothpicks are added to the stage 1 design to get the stage 2 design? How many are added to the stage 2 design to get the stage 3 design? How many are added to the stage 3 design to get the stage 4 design? How many toothpicks are added to the stage $(n - 1)$ design to get the stage n design?

10. Write a recursive formula that tells the *total* number of toothpicks used in the stage n design.

For each sequence, find the first 4 terms. Then write an explicit formula. *(Section 10.3)*

11. $t_1 = -22; t_n = t_{n-1} + 5$

12. $t_1 = 200; t_n = (1.05)t_{n-1}$

For Exercises 13 and 14, write a recursive formula for each sequence. *(Section 10.3)*

13. 1.00, 1.11, 1.22, 1.33, …   **14.** 100, 80, 64, 51.2, …

15. Write $\displaystyle\sum_{n=1}^{6} \frac{n^2}{100}$ in expanded form. *(Section 10.4)*

16. Write the sequence $3 + 5 + 7 + \cdots + 131$ in sigma notation. *(Section 10.4)*

17. Which of the series in Exercises 15 and 16 are arithmetic? Find the sum(s) of the arithmetic series. *(Section 10.4)*

18. The first term of a finite arithmetic series is 8. The last term is –125. The sum is –1170. How many terms are in the series? *(Section 10.4)*

For Exercises 19–22, find the sum of each series, if the sum exists. If the series does not have a sum, explain why. *(Section 10.5)*

19. $48 + 24 + 12 + \cdots + 1.5$ **20.** $\displaystyle\sum_{n=1}^{14} 0.0001(1.6)^n$

21. $\displaystyle\sum_{n=1}^{\infty} 20\left(\frac{6}{5}\right)^n$ **22.** $\displaystyle\sum_{n=1}^{\infty} 6\left(\frac{2}{3}\right)^n$

23. Express the repeating decimal 3.434343… as a fraction. *(Section 10.6)*

SPIRAL REVIEW   **Chapters 1–10**

Solve each equation. Round your answers to the nearest thousandth.

1. $\dfrac{y}{15} = \dfrac{6}{y+1}$ **2.** $\dfrac{2}{w} = 1 - \dfrac{1}{w-2}$ **3.** $\log_2 x + \log_2 (x - 6) = 4$

4. $240 = 3 + 10e^{4t}$ **5.** $5^x = 125^{0.2}$ **6.** $\log (x + 2) = 3 + \log x$

For Exercises 7–9, use the points $A(3, 5)$ and $B(6, 12)$.

7. Write an equation of the line passing through points A and B.

8. Write an equation for the exponential function passing through points A and B.

9. Write an equation for the quadratic function passing through point A, point B, and the origin.

For Exercises 10–12, use matrices A and B below.

$$A = \begin{bmatrix} 6 & 3 \\ 0 & 4 \\ -2 & 0 \end{bmatrix} \qquad B = \begin{bmatrix} 1 & 0 \\ 5 & 1 \\ 0 & -12 \end{bmatrix}$$

10. $A - B$ **11.** $\dfrac{1}{2}A$ **12.** $-B$

Distances, Midpoints, and Lines

Learn how to . . .

- find the distance between two points in a coordinate plane

- find the coordinates of the midpoint of a line segment in a coordinate plane

So you can . . .

- solve problems involving distances

Application

To determine the distance between the attractions at a large amusement park, a coordinate grid is superimposed onto a scale drawing of the entire park so that the distance formula can be used.

UNDERSTANDING THE MAIN IDEAS

The distance and midpoint formulas

The distance formula, $d = \sqrt{(x_2 - x_1)^2 + (y_2 - y_1)^2}$, is used to find the distance between two points, (x_1, y_1) and (x_2, y_2), located on a coordinate plane. The midpoint formula, $\left(\dfrac{x_1 + x_2}{2}, \dfrac{y_1 + y_2}{2} \right)$, is used to find the midpoint of the segment with endpoints (x_1, y_1) and (x_2, y_2).

Example 1

For the pair of points, (1, 2) and (5, −1), find:

  a. the distance between the points.

  b. the coordinates of the midpoint of the line segment connecting the points.

■ **Solution** ■

a. Let $(x_2, y_2) = (5, -1)$ and $(x_1, y_1) = (1, 2)$.

$$d = \sqrt{(x_2 - x_1)^2 + (y_2 - y_1)^2}$$
$$= \sqrt{(5 - 1)^2 + (-1 - 2)^2}$$
$$= \sqrt{4^2 + (-3)^2}$$
$$= \sqrt{16 + 9}$$
$$= \sqrt{25}$$
$$= 5$$

The distance between the points $(1, 2)$ and $(5, -1)$ is 5.

b. The midpoint of the segment between $(1, 2)$ and $(5, -1)$ is:

$$\left(\frac{x_1 + x_2}{2}, \frac{y_1 + y_2}{2}\right) = \left(\frac{1 + 5}{2}, \frac{2 + (-1)}{2}\right)$$
$$= \left(\frac{6}{2}, \frac{1}{2}\right)$$
$$= \left(3, \frac{1}{2}\right)$$

For each pair of points, find:

a. the distance between the points.

b. the coordinates of the midpoint of the line segment connecting the points.

1. $(2, -2), (7, 3)$ **2.** $(-3, 5), (7, 3)$ **3.** $(-7, 1), (-4, 5)$

4. $(-1, -4), (4, -3)$ **5.** $(-2, 3), (4, 0)$ **6.** $(-2, -3), (3, 6)$

You can also use the distance formula to find a particular point that is a specified distance from a given point.

Example 2

Find the value(s) of k so that the points $(-2, 4)$ and $(3, k)$ are 13 units apart.

■ Solution ■

Use the distance formula $d = \sqrt{(x_2 - x_1)^2 + (y_2 - y_1)^2}$ with $(x_1, y_1) = (-2, 4)$, $(x_2, y_2) = (3, k)$, and $d = 13$.

$$13 = \sqrt{(3 - (-2))^2 + (k - 4)^2}$$

$$13 = \sqrt{5^2 + (k - 4)^2}$$

$$169 = 5^2 + (k - 4)^2 \qquad \leftarrow \text{Square both sides of the equation.}$$

$$169 = 25 + (k - 4)^2$$

$$144 = (k - 4)^2 \qquad \leftarrow \text{Take the square root of both sides.}$$

$$\pm 12 = k - 4$$

$$k - 4 = 12 \quad \text{or} \quad k - 4 = -12$$

$$k = 16 \quad \text{or} \quad k = -8$$

The two points are 13 units apart when the value of k is -8 or 16.

Find a value of k so that the given points are n units apart.

7. $(5, k)$, $(2, 6)$; $n = 5$ **8.** $(6, 7)$, $(k, 11)$; $n = 5$

9. $(3.5, k)$, $(-1.3, 2.8)$; $n = 5$ **10.** $(-2, 3)$, $(6, k)$; $n = 17$

11. $(k, 4)$, $(7, 19)$; $n = 17$ **12.** $(k, -2)$, $(7, 10)$; $n = 13$

Parallel and perpendicular lines

Two lines are parallel if and only if they have the same slope. Two lines are perpendicular if and only if their slopes are negative reciprocals of each other.

Example 3

Find an equation of the perpendicular bisector of the line segment connecting the points $(1, 2)$ and $(5, -1)$.

■ Solution ■

A bisector passes through the midpoint of a segment. So the perpendicular bisector of the segment connecting $(1, 2)$ and $(5, -1)$ passes through the point $\left(3, \dfrac{1}{2}\right)$, the midpoint found in Example 1.

The slope m of the line through $(1, 2)$ and $(5, -1)$ is $\dfrac{-1-2}{5-1}$, or $-\dfrac{3}{4}$. So the slope of a line that is perpendicular to this line is the negative reciprocal of $-\dfrac{3}{4}$, or $\dfrac{4}{3}$.

(Solution continues on next page.)

■ **Solution** ■ *(continued)*

Now find the line with slope of $\frac{4}{3}$ that passes through the point $\left(3, \frac{1}{2}\right)$.

$$y = ax + b$$

$$\frac{1}{2} = \frac{4}{3}(3) + b$$

$$\frac{1}{2} = 4 + b$$

$$-\frac{7}{2} = b$$

The equation of the perpendicular bisector is $y = \frac{4}{3}x - \frac{7}{2}$.

For each pair of points, find an equation of the perpendicular bisector of the line segment connecting the points.

13. $(4, 5), (-1, 3)$ **14.** $(5, 1), (-4, -6)$ **15.** $(-1, 1), (5, -1)$

....
Spiral Review

For Exercises 16 and 17, write each geometric series in sigma notation. Find the sum of each series, if the sum exists. If a series does not have a sum, explain why not. *(Section 10.5)*

16. $12 + (-6) + 3 + (-1.5) + \cdots$ **17.** $4 + (-8) + 16 + (-32) + \cdots + 1024$

18. Graph the feasible region for the system of inequalities and find the minimum cost for the cost function $C = 200x + 50y$. *(Section 7.6)*

$x + y \le 20$
$y \ge x$
$90x + 30y \ge 720$

Describe the graph of each function. Make a sketch of each graph.
(Section 5.2)

19. $y = 3(x - 2)^2 + 1$ **20.** $y = -2(x + 2)^2 - 3$

Parabolas

Learn how to . . .

- find the focus and directrix of a parabola

- write an equation of a parabola

So you can . . .

- solve problems involving parabolas

Application

Because of the way that sound waves focus when two parabolic dishes are positioned facing each other, a whisper at the focus of one dish can be heard at the focus of the other dish regardless of the distance between them.

Terms to Know

Example / Illustration

Parabola (p. 521) the set of all points in a plane that are the same distance from a fixed point and a fixed line	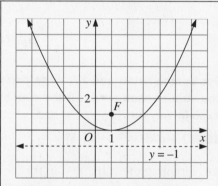
Focus (p. 521) the fixed point that defines a parabola	
Directrix (p. 521) the fixed line that defines a parabola	The focus of the parabola is the point $F(1, 1)$, the directrix is the line $y = -1$, and the vertex is $(1, 0)$.
Vertex (p. 521) the point on a parabola that lies halfway between the focus and the directrix	

UNDERSTANDING THE MAIN IDEAS

The general equation of a parabola that opens up or down is $y - k = \dfrac{1}{4c}(x - h)^2$,

where the vertex is (h, k), the focus is $(h, k + c)$, the directrix is $y = k - c$, and the line of symmetry of the parabola is $x = h$. The parabola opens up if $c > 0$ and opens down if $c < 0$.

The general equation of a parabola that opens right or left is $x - h = \frac{1}{4c}(y - k)^2$, where the vertex is (h, k), the focus is $(h + c, k)$, the directrix is $x = h - c$, and the line of symmetry of the parabola is $y = k$. The parabola opens right if $c > 0$ and opens left if $c < 0$.

Example 1

Find an equation of the parabola with the given characteristics.

  a. vertex $(0, 0)$; focus $(0, -3)$

  b. focus $(-2, 2)$; directrix $x = 2$

▪ Solution ▪

  a. Since the focus is *below* the vertex, this parabola opens down and its equation has the form $y - k = \frac{1}{4c}(x - h)^2$ with $c < 0$. The given vertex indicates that $h = 0$ and $k = 0$. Since the parabola opens down, the focus $(0, -3)$ is of the form $(h, k + c)$. Therefore,

$$k + c = -3$$
$$0 + c = -3$$
$$c = -3$$

  The equation of the parabola is $y - 0 = \frac{1}{4(-3)}(x - 0)^2$, or $y = -\frac{1}{12}x^2$.

  b. Since the directrix is to the *right* of the focus, this parabola opens to the left and its equation has the form $x - h = \frac{1}{4c}(y - k)^2$ with $c < 0$. The vertex of the parabola is located halfway between the focus and the directrix. That is, it is the midpoint of the segment connecting $(-2, 2)$ and $(2, 2)$. So the vertex is

$$\left(\frac{-2 + 2}{2}, \frac{2 + 2}{2}\right) = (0, 2)$$

  Thus $h = 0$ and $k = 2$.

  The focus $(-2, 2)$ has the form $(h + c, k)$. Therefore,

$$h + c = -2$$
$$0 + c = -2$$
$$c = -2$$

  The equation of the parabola is $x - 0 = \frac{1}{4(-2)}(y - 2)^2$, or $x = -\frac{1}{8}(y - 2)^2$.

For Exercises 1–6:

  a. Find an equation of the parabola with the given characteristics.

  b. Graph your equation from part (a).

  1. vertex $(0, 0)$; directrix $y = 1$ **2.** vertex $(2, 0)$; directrix $y = -2$

  3. vertex $(0, 0)$; focus $(0, -2)$ **4.** vertex $(0, 0)$; directrix $y = -2$

  5. focus $(3, 2)$; vertex $(1, 2)$ **6.** vertex $(-1, 1)$; directrix $x = 2$

If you know the equation of a parabola, you can use it to identify the distinguishing characteristics of the parabola. If the equation is not given in standard form, you will need to rewrite it first.

Example 2

Name the vertex, focus, and directrix of a parabola with each equation.

  a. $y = -\dfrac{1}{6}(x - 4)^2$ **b.** $x = y^2 - 8$

■ Solution ■

  a. The equation has the form $y - k = \dfrac{1}{4c}(x - h)^2$ with $h = 4$ and $k = 0$. So the vertex (h, k) is $(4, 0)$.

  To identify the focus and directrix, you must find the value of c.

  $$\frac{1}{4c} = -\frac{1}{6}$$
  $$4c = -6$$
  $$c = -1.5$$

  The focus $(h, k + c)$ is $(4, 0 - 1.5)$, or $(4, -1.5)$.

  The directrix has equation $y = k - c$: $y = 0 - (-1.5)$, or $y = 1.5$.

  b. Rewrite the equation in the form $x - h = \dfrac{1}{4c}(y - k)^2$.

  $$x = (y - 0)^2 - 8$$
  $$x + 8 = (y - 0)^2$$
  $$x - (-8) = (y - 0)^2$$

  From the equation, the vertex (h, k) is $(-8, 0)$.

  To identify the focus and directrix, you must find the value of c.

  $$\frac{1}{4c} = 1$$
  $$1 = 4c$$
  $$c = 0.25$$

  So the focus $(h + c, k)$ is $(-8 + 0.25, 0)$, or $(-7.75, 0)$, and the directrix $x = h - c$ is $x = -8 - 0.25$, or $x = -8.25$.

Name the vertex, focus, and directrix of a parabola with the given equation. Sketch the parabola with its focus and directrix.

7. $y = -\frac{1}{8}(x - 2)^2$ **8.** $x = \frac{1}{2}y^2$ **9.** $x = -\frac{1}{8}(y + 2)^2$

10. $y = \frac{1}{2}(x - 1)^2$ **11.** $x - 2 = y^2$ **12.** $x = \frac{1}{12}(y + 2)^2$

....
Spiral Review

For each pair of points, find:

a. the distance between the points.

b. the coordinates of the midpoint of the line segment connecting the points. *(Section 11.1)*

13. $(-6, 1), (2, -3)$ **14.** $(3, 0), (-4, -1)$ **15.** $(5, 2), (3, -4)$

Solve each system of equations. *(Section 7.1)*

16. $3x + 2y = 1$
$x + y = 1$

17. $3x - y = 12$
$2x + 3y = 74$

18. $4x + 2y = 1$
$-6x + 4y = 9$

Circles

Learn how to . . .

- write an equation
  of a circle

- graph an equation
  of a circle

So you can . . .

- locate points on a
  circle

Application

A cellular telephone network uses beacons to transmit calls. Each beacon has a range that is a circular area. The radius of the circle determines the area that each beacon covers. The circles must overlap, or there will be locations where there is no service available.

Terms to Know

Example / Illustration

Circle (p. 528) the set of all points in a plane that are the same distance from a point in the plane	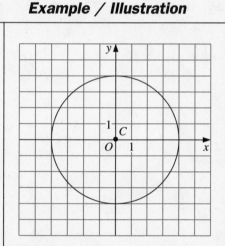
Center (p. 528) the point from which the measure to all points on a circle is made	The center of the circle above is $C(0, 0)$.
Radius (p. 528) the distance between the center of a circle and any point on the circle	The radius of the circle above is 4 units.

UNDERSTANDING THE MAIN IDEAS

The standard form for the equation of a circle with center $C(h, k)$ and radius r is
$(x - h)^2 + (y - k)^2 = r^2$.

Example 1

For each equation of a circle, identify the center and the radius. Then graph the equation.

a. $(x - 1)^2 + (y + 2)^2 = 14$ 　　　　 **b.** $x^2 + y^2 + 6y + 9 = 2$

■ Solution ■

a. Rewrite the given equation in the form $(x - h)^2 + (y - k)^2 = r^2$ and then identify the center (h, k) and the radius r.

$$(x - 1)^2 + (y - (-2))^2 = (\sqrt{14})^2$$

So the center of the circle is $(1, -2)$ and the radius is $\sqrt{14} \approx 3.74$.

The graph of the equation is the circle shown at the right.

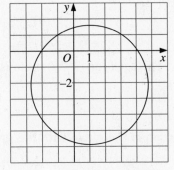

b. To rewrite the given equation in the form $(x - h)^2 + (y - k)^2 = r^2$, notice that $y^2 + 6y + 9 = (y + 3)^2$.

$$(x - 0)^2 + (y - (-3))^2 = (\sqrt{2})^2$$

So the center of the circle is $(0, -3)$ and the radius is $\sqrt{2} \approx 1.41$.

The graph of the equation is the circle shown at the right.

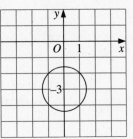

For each equation of a circle, identify the center and the radius. Then graph the equation.

1. $(x - 2)^2 + (y - 2)^2 = 4$ 　　　　 **2.** $(x - 1)^2 + y^2 = 25$

3. $(x + 2)^2 + (y - 5)^2 = 36$ 　　　　 **4.** $(x + 1)^2 + (y + 2)^2 = 9$

5. $x^2 + y^2 + 10y + 25 = 15$ 　　　　 **6.** $x^2 - 4x + 4 + y^2 = 6$

To write the equation of a circle when you know the center and radius, simply insert the known values of h, k, and r into the standard equation $(x - h)^2 + (y - k)^2 = r^2$.

Example 2

Write an equation of the circle with the given center C and radius r.

a. $C(2, -3)$; $r = 4$ 　　　　 **b.** $C(0, -\sqrt{3})$; $r = \sqrt{5}$

■ Solution ■

a. Use $(x - h)^2 + (y - k)^2 = r^2$ with $h = 2$, $k = -3$, and $r = 4$.

$$(x - 2)^2 + (y - (-3))^2 = 4^2$$
$$(x - 2)^2 + (y + 3)^2 = 16$$

The equation of the circle is $(x - 2)^2 + (y + 3)^2 = 16$.

b. $(x - h)^2 + (y - k)^2 = r^2$ $\leftarrow h = 0$, $k = -\sqrt{3}$, and $r = \sqrt{5}$

$$(x - 0)^2 + (y - (-\sqrt{3}))^2 = (\sqrt{5})^2$$
$$x^2 + (y + \sqrt{3})^2 = 5$$

The equation of the circle is $x^2 + (y + \sqrt{3})^2 = 5$.

Write an equation of the circle with the given center C and radius r.

7. $C(2, 0)$; $r = 4$ **8.** $C(-2, -1)$; $r = 2$ **9.** $C(0, -3)$; $r = 1$

10. $C(1, -1)$; $r = 9$ **11.** $C(2, \sqrt{3})$; $r = 11$ **12.** $C(-3, -\sqrt{3})$; $r = \sqrt{7}$

13. $C(2, -3)$; $r = \sqrt{6}$ **14.** $C(-1, 0)$; $r = 5\sqrt{3}$ **15.** $C(a, b)$; $r = 2c$

....

Spiral Review

Find an equation of each parabola with the given focus and directrix. Sketch the parabola with its focus and directrix. *(Section 11.2)*

16. focus $(0, -3)$
   directrix $y = 3$

17. focus $(-2, 2)$
   directrix $x = 2$

Graph each function. State the domain and range. *(Section 8.1)*

18. $y = \sqrt{x - 1}$

19. $y = \sqrt{x + 2} + 1$

Ellipses

Learn how to . . .

- write an equation of an ellipse

- graph an equation of an ellipse

So you can . . .

- describe an elliptical object

Application

In Statuary Hall in the United States Capitol, there is an elliptical chamber in which a person whispering at one focus can easily be heard by another person standing at the other focus. The hall is 46 ft wide and 96 ft long.

Terms to Know	**Example / Illustration**
Ellipse (p. 536) the set of all points in a plane such that the sum of the distances between any point and two fixed points is constant	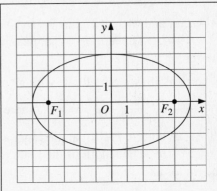 The equation of the ellipse above is $\dfrac{x^2}{25} + \dfrac{y^2}{9} = 1$.
Foci (p. 536) the two fixed points that define an ellipse (*Foci* is the plural of *focus.*)	The foci of the ellipse above are $(-4, 0)$ and $(4, 0)$.
Center (p. 536) the midpoint of the line segment connecting the foci	The center of the ellipse above is $(0, 0)$.
Vertices (p. 536) the points of intersection of the ellipse and the line containing the foci	The vertices of the ellipse above are $(-5, 0)$ and $(5, 0)$.

Major axis (p. 536) the line segment that contains the foci and has the vertices at its endpoints	The major axis of the ellipse shown on the previous page has endpoints (–5, 0) and (5, 0). Its length is 2(5), or 10 units.
Minor axis (p. 536) the line segment through the center and perpendicular to the major axis, with endpoints on the ellipse	The minor axis of the ellipse shown on the previous page has endpoints (0, –3) and (0, 3). Its length is 2(3), or 6 units.

UNDERSTANDING THE MAIN IDEAS

A horizontal ellipse (an ellipse with a horizontal major axis) has equation $\frac{(x-h)^2}{a^2} + \frac{(y-k)^2}{b^2} = 1$, $a > b > 0$, where its center is (h, k), its vertices are $(h - a, k)$ and $(h + a, k)$, and its foci are $(h - c, k)$ and $(h + c, k)$ where $c > 0$ and $c^2 = a^2 - b^2$. The length of the major axis of the ellipse is $2a$ and the length of the minor axis is $2b$.

A vertical ellipse (an ellipse with a vertical major axis) has equation $\frac{(x-h)^2}{b^2} + \frac{(y-k)^2}{a^2} = 1$, $a > b > 0$, where its center is (h, k), its vertices are $(h, k - a)$ and $(h, k + a)$, and its foci are $(h, k - c)$ and $(h, k + c)$ where $c > 0$ and $c^2 = a^2 - b^2$. The length of the major axis of the ellipse is $2a$ and the length of the minor axis is $2b$.

Example

Identify the center, the vertices, and the foci of each ellipse. Tell whether the ellipse is horizontal or vertical. Then graph the equation.

a. $\frac{(x-2)^2}{25} + \frac{(y-3)^2}{16} = 1$ **b.** $\frac{(x-4)^2}{16} + \frac{(y+1)^2}{36} = 1$

Solution

a. Since the denominator of the term involving x is greater than the denominator of the term involving y, this ellipse is *horizontal*. So its major axis is horizontal.

From the equation, $h = 2$, $k = 3$, $a = \sqrt{25} = 5$, and $b = \sqrt{16} = 4$. You can now use the values of h, k, and a to find the center and the vertices.

(Solution continues on next page.)

■ **Solution** ■ *(continued)*

The center (h, k) is $(2, 3)$. The vertices $(h - a, k)$ and $(h + a, k)$ are $(2 - 5, 3) = (-3, 3)$ and $(2 + 5, 3) = (7, 3)$.

To find the foci, you need to know the value of c. Using $a = 5$ and $b = 4$,

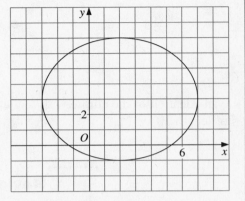

$$c^2 = 5^2 - 4^2$$
$$= 25 - 16$$
$$= 9$$
$$c = 3 \qquad \leftarrow \text{Remember, } c > 0.$$

The foci $(h - c, k)$ and $(h + c, k)$ are $(2 - 3, 3) = (-1, 3)$ and $(2 + 3, 3) = (5, 3)$.

The graph is shown at the right.

b. Since the denominator of the term involving y is greater than the denominator of the term involving x, this ellipse is *vertical*. So its major axis is vertical.

From the equation, $h = 4$, $k = -1$, $a = \sqrt{36} = 6$, and $b = \sqrt{16} = 4$. You can now use the values of h, k, and a to find the center and the vertices.

The center (h, k) is $(4, -1)$. The vertices $(h, k - a)$ and $(h, k + a)$ are $(4, -1 - 6) = (4, -7)$ and $(4, -1 + 6) = (4, 5)$.

To find the foci, you need to know the value of c. Using $a = 6$ and $b = 4$,

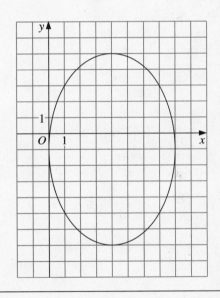

$$c^2 = 6^2 - 4^2$$
$$= 36 - 16$$
$$= 20$$
$$c = \sqrt{20}, \text{ or } 2\sqrt{5}$$

The foci $(h, k - c)$ and $(h, k + c)$ are $(4, -1 - 2\sqrt{5})$ and $(4, -1 + 2\sqrt{5})$, or about $(4, -5.47)$ and about $(4, 3.47)$.

The graph is shown at the right.

For Exercises 1–7, identify the center, the vertices, and the foci of each ellipse with the given equation. Tell whether the ellipse is *horizontal* or *vertical*. Then graph the equation.

1. $\dfrac{x^2}{25} + \dfrac{y^2}{9} = 1$

2. $\dfrac{x^2}{25} + \dfrac{y^2}{49} = 1$

3. $\dfrac{(x-2)^2}{16} + \dfrac{(y-3)^2}{9} = 1$

4. $\dfrac{(x-3)^2}{4} + \dfrac{(y-2)^2}{1} = 1$

5. $\dfrac{(x-1)^2}{4} + \dfrac{(y+2)^2}{16} = 1$

6. $\dfrac{(x-4)^2}{2} + \dfrac{(y+3)^2}{3} = 1$

7. $4x^2 + 9y^2 = 36$

8. Write an equation of the ellipse with foci at (–4, 2) and (4, 2), and major axis of length 10.

· · · · · · · · · · · · · · · · · · · ·
Spiral Review

Write an equation of the circle with the given center *C* and radius *r*. Then sketch its graph. *(Section 11.3)*

9. $C(2, -3);\ r = 4$

10. $C(0, -\sqrt{3});\ r = \sqrt{5}$

Find the slope-intercept equation of each line. *(Section 2.2)*

11.

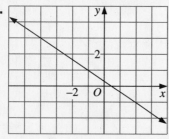

12.

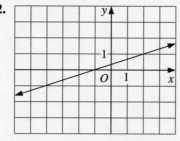

Hyperbolas

Application

A ray of light directed toward one focus of a hyperbolic mirror is reflected toward the other focus. This property is used in telescopes to focus light.

Terms to Know

Example / Illustration

Hyperbola (p. 543) the set of all points in a plane such that the difference of the distances between a point and each of two foci is constant	The equation of the hyperbola below is $\dfrac{x^2}{4} - \dfrac{y^2}{9} = 1$. 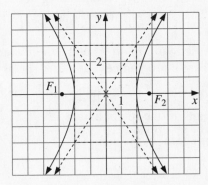
Foci (p. 543) the two fixed points that define a hyperbola	The foci of the hyperbola above are $(-2\sqrt{2}, 0)$ and $(2\sqrt{2}, 0)$.
Center (p. 543) the midpoint of the line segment that connects the foci	The center of the hyperbola above is $(0, 0)$.
Vertices (p. 543) the points of intersection of the hyperbola and the line containing the foci	The vertices of the hyperbola above are $(-2, 0)$ and $(2, 0)$.

Major axis (p. 543) the line segment that has the vertices as its endpoints	The major axis of the hyperbola shown on the previous page has endpoints (–2, 0) and (2, 0). Its length is 2(2), or 4 units.
Minor axis (p. 543) the line segment through the center and perpendicular to the major axis	The minor axis of the hyperbola shown on the previous page has endpoints (0, –3) and (0, 3). Its length is 2(3), or 6 units.

UNDERSTANDING THE MAIN IDEAS

Many parts of the standard form of the equation of a hyperbola are similar to the corresponding parts of the standard equation of an ellipse. Look for the negative sign to distinguish between a hyperbola equation and an ellipse equation. When graphing the equation of a hyperbola, the order of the terms should be noted in order to determine whether the hyperbola is horizontal or vertical. Unlike an ellipse, the graph of a hyperbola has a pair of asymptotes associated with it.

A horizontal hyperbola (a hyperbola with a horizontal major axis) has equation $\dfrac{(x-h)^2}{a^2} - \dfrac{(y-k)^2}{b^2} = 1$, $a > 0$ and $b > 0$, where its center is (h, k), its vertices are $(h - a, k)$ and $(h + a, k)$, and its foci are $(h - c, k)$ and $(h + c, k)$ where $c > 0$ and $c^2 = a^2 + b^2$. The asymptotes of the graph of this hyperbola are $y = \pm \dfrac{b}{a}(x - h) + k$. The length of the major axis of the hyperbola is $2a$ and the length of the minor axis is $2b$.

A vertical hyperbola (a hyperbola with a vertical major axis) has equation $\dfrac{(y-k)^2}{a^2} - \dfrac{(x-h)^2}{b^2} = 1$, $a > 0$ and $b > 0$, where its center is (h, k), its vertices are $(h, k - a)$ and $(h, k + a)$, and its foci are $(h, k - c)$ and $(h, k + c)$ where $c > 0$ and $c^2 = a^2 + b^2$. The asymptotes of the graph of this hyperbola are $y = \pm \dfrac{a}{b}(x - h) + k$. The length of the major axis of the hyperbola is $2a$ and the length of the minor axis is $2b$.

Example

Identify the center, the vertices, and the foci of each hyperbola with the given equation. Tell whether the hyperbola is *horizontal* or *vertical*. Find equations of the asymptotes. Then graph the equation.

a. $\dfrac{(x+3)^2}{16} - \dfrac{(y-2)^2}{9} = 1$ **b.** $\dfrac{y^2}{36} - \dfrac{(x-2)^2}{64} = 1$

■ Solution ■

a. Since the term involving y is subtracted from the term involving x, this hyperbola is *horizontal*. So its major axis is horizontal.

From the equation, $h = -3$, $k = 2$, $a = \sqrt{16} = 4$, and $b = \sqrt{9} = 3$. You can now use the values of h, k, a, and b to find the center, the vertices, and the asymptotes.

The center (h, k) is $(-3, 2)$. The vertices $(h - a, k)$ and $(h + a, k)$ are $(-3 - 4, 2) =$ $(-7, 2)$ and $(-3 + 4, 2) = (1, 2)$. The asymptotes $y = \pm \dfrac{b}{a}(x - h) + k$ are $y = \dfrac{3}{4}(x + 3) + 2$ and $y = -\dfrac{3}{4}(x + 3) + 2$.

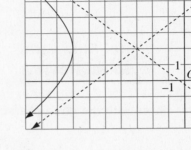

To find the foci, you need to know the value of c. Using $a = 4$ and $b = 3$,

$$c^2 = 4^2 + 3^2$$
$$= 16 + 9$$
$$= 25$$
$$c = 5 \qquad \leftarrow \text{Remember, } c > 0.$$

The foci $(h - c, k)$ and $(h + c, k)$ are $(-3 - 5, 2) = (-8, 2)$ and $(-3 + 5, 2)$ $= (2, 2)$.

The graph is shown at the right.

b. Since the term involving x is subtracted from the term involving y, this hyperbola is *vertical*. So its major axis is vertical.

From the equation, $h = 2$, $k = 0$, $a = \sqrt{36} = 6$, and $b = \sqrt{64} = 8$. You can now use the values of h, k, a, and b to find the center, the vertices, and the asymptotes.

The center (h, k) is $(2, 0)$. The vertices $(h, k - a)$ and $(h, k + a)$ are $(2, 0 - 6) =$ $(2, -6)$ and $(2, 0 + 6) = (2, 6)$. The asymptotes $y = \pm \dfrac{a}{b}(x - h) + k$ are $y = \dfrac{3}{4}(x - 2)$ and $y = -\dfrac{3}{4}(x - 2)$.

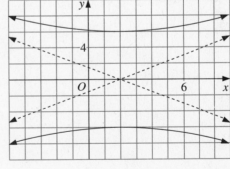

To find the foci, you need to know the value of c. Using $a = 6$ and $b = 8$,

$$c^2 = 6^2 + 8^2$$
$$= 36 + 64$$
$$= 100$$
$$c = 10$$

The foci $(h, k - c)$ and $(h, k + c)$ are $(2, 0 - 10) = (2, -10)$ and $(2, 0 + 10)$ $= (2, 10)$.

The graph is shown at the right.

For Exercises 1–7, identify the center, the vertices, and the foci of each hyperbola with the given equation. Tell whether the hyperbola is *horizontal* or *vertical*. Find equations of the asymptotes. Then graph the equation.

1. $\dfrac{x^2}{9} - \dfrac{y^2}{16} = 1$

2. $\dfrac{y^2}{36} - \dfrac{x^2}{64} = 1$

3. $\dfrac{(x-5)^2}{4} - \dfrac{y^2}{9} = 1$

4. $\dfrac{x^2}{49} - \dfrac{(y-6)^2}{25} = 1$

5. $\dfrac{(y-1)^2}{4} - \dfrac{(x-6)^2}{49} = 1$

6. $\dfrac{(x-2)^2}{4} - \dfrac{(y-2)^2}{9} = 1$

7. $4y^2 - 16x^2 = 64$

8. **Mathematics Journal** Explain what the values of h, k, a, b, and c tell you about the graph of an ellipse and what they tell you about the graph of a hyperbola. How are they similar and how are they different for these two different curves?

················
Spiral Review

Find an equation of the ellipse with the given characteristics. Then sketch the graph. *(Section 11.4)*

9. center $(-2, 3)$; $a = 5$; $b = 4$; major axis is vertical.

10. center $(6, 3)$; vertex $(1, 3)$; focus $(2, 3)$

Write the equation of each function in vertex form and in intercept form. *(Sections 5.4 and 5.5)*

11. $y = x^2 - 2x - 8$

12. $y = -2x^2 - 7x + 15$

Identifying Conics

Learn how to . . .

- find conics by taking a cross section of a double cone

So you can . . .

- identify various conics from their equations

Application

The orbits of many celestial bodies (such as Halley's comet) are elliptical, some being very close to circular. Some comets have a hyperbolic path, with our sun at a focus. These comets pass by the sun only once.

Terms to Know

Example / Illustration

Terms to Know	Example / Illustration
Conic section (p. 550) a curve produced by cutting across a double cone with a plane	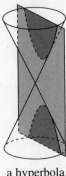 a hyperbola
Degenerate conic (p. 551) a conic produced when a plane intersects a double cone at its vertex	The degenerate conic $4x^2 - y^2 = 0$ is a pair of intersecting lines, while the degenerate conic $4x^2 + y^2 = 0$ is simply a point.

UNDERSTANDING THE MAIN IDEAS

The equation $Ax^2 + Bxy + Cy^2 + Dx + Ey + F = 0$ has a graph that is either a conic section or a degenerate conic. Completing the square in x and/or y will produce a form of the equation that can be used to identify the graph as a circle, a parabola, an ellipse, a hyperbola, a point, or a pair of intersecting lines.

Example 1

For each equation of a conic, rewrite the equation in standard form, identify the conic, and state its important characteristics.

a. $y^2 - 6x - 4y + 22 = 0$ **b.** $9x^2 + 16y^2 - 36x - 32y - 92 = 0$

■ Solution ■

a. Since there is a y^2-term, complete the square in y.

$$y^2 - 6x - 4y + 22 = 0$$
$$y^2 - 4y - 6x + 22 = 0 \quad \leftarrow \text{Regroup terms.}$$
$$(y^2 - 4y + \underline{?}) - 6x + 22 - \underline{?} = 0 \quad \leftarrow \text{Complete the square in } y.$$
$$(y^2 - 4y + 4) - 6x + 22 - 4 = 0$$
$$(y - 2)^2 - 6x + 18 = 0$$
$$(y - 2)^2 - 6(x - 3) = 0$$
$$x - 3 = \frac{1}{6}(y - 2)^2$$

This is the equation of a parabola that opens to the right, with vertex (3, 2). The focus is $\left(\frac{11}{2}, 2\right)$ and the directrix is $x = \frac{3}{2}$.

b. Since there is an x^2-term and a y^2-term, complete the square in both x and y.

$$9x^2 + 16y^2 - 36x - 32y - 92 = 0$$
$$9x^2 - 36x + 16y^2 - 32y - 92 = 0$$
$$9(x^2 - 4x + \underline{?}) + 16(y^2 - 2y + \underline{?}) - 92 - \underline{?} - \underline{?} = 0$$
$$9(x^2 - 4x + 4) + 16(y^2 - 2y + 1) - 92 - 9(4) - 16(1) = 0$$
$$9(x - 2)^2 + 16(y - 1)^2 - 144 = 0$$
$$9(x - 2)^2 + 16(y - 1)^2 = 144$$
$$\frac{(x - 2)^2}{16} + \frac{(y - 1)^2}{9} = 1$$

This is the equation of an ellipse with center (2, 1), a horizontal major axis, vertices (6, 1) and (–2, 1), and foci $(2 - \sqrt{7}, 1)$ and $(2 + \sqrt{7}, 1)$. The endpoints of the minor axis are (2, 4) and (2, –2).

For each equation of a conic, rewrite the equation in standard form, identify the conic, and state its important characteristics.

1. $2x^2 + 2y^2 - 8x + 5 = 0$

2. $x^2 + 4y^2 - 6x - 16y + 21 = 0$

3. $y^2 - 6y + 10x - 1 = 0$

4. $9x^2 - 4y^2 - 90x + 189 = 0$

5. $x^2 - y^2 + 6x + 4y + 14 = 0$

6. $16x^2 + 4y^2 - 32x + 16y - 32 = 0$

Degenerate conics

There are three types of degenerate conics. The degenerate case for an ellipse or circle is a single point, the degenerate case for a parabola is a line, and the degenerate case for a hyperbola is a pair of intersecting lines.

Example 2

Identify and graph each degenerate conic.

a. $3x^2 + 3y^2 = 0$

b. $x^2 - y^2 + 2x + 6y - 8 = 0$

■ Solution ■

a. $3x^2 + 3y^2 = 0$
$3(x^2 + y^2) = 0$
$x^2 + y^2 = 0$

The equation $x^2 + y^2 = 0$ has only one solution,
$x = 0$ and $y = 0$. The graph of this equation is the point
$(0, 0)$. This is the degenerate case for an ellipse or a circle.

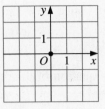

b.
$$x^2 - y^2 + 2x + 6y - 8 = 0$$
$$(x^2 + 2x + \underline{?}) - (y^2 - 6y + \underline{?}) - 8 - \underline{?} + \underline{?} = 0$$
$$(x^2 + 2x + 1) - (y^2 - 6y + 9) - 8 - 1 + 9 = 0$$
$$(x + 1)^2 - (y - 3)^2 + 0 = 0$$
$$(x + 1)^2 - (y - 3)^2 = 0$$
$$(y - 3)^2 = (x + 1)^2$$
$$y - 3 = \pm(x + 1)$$
$$y = 3 \pm (x + 1)$$

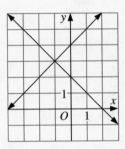

The graph of the equation is the pair of intersecting lines
$y = x + 4$ and $y = -x + 2$. This is the degenerate case for
a hyperbola.

Identify and graph each equation of a degenerate conic.

7. $9x^2 - y^2 = 0$

8. $x^2 + y^2 = 0$

9. $x^2 + y^2 + 6x - 4y + 13 = 0$

10. $x^2 + 2y^2 - 2x + 1 = 0$

· · · · · · · · · · · · · · · · · · · ·
Spiral Review

**Identify the center, the vertices, and the foci of each hyperbola. Tell whether
the hyperbola is *horizontal* or *vertical*. Find the equations of the asymptotes.
Then graph the equation.** *(Section 11.5)*

11. $\dfrac{(x + 3)^2}{16} - \dfrac{(y - 2)^2}{9} = 1$

12. $\dfrac{y^2}{36} - \dfrac{(x - 2)^2}{64} = 1$

Graph each system of inequalities. *(Section 7.5)*

13. $y \geq 3x$
$y \leq -2x + 1$

14. $y \leq 3x - 3$
$y > -0.5x + 2$

15. $y \geq -x + 4$
$y \leq 2x + 2$

Chapter 11 Review ·············

Complete these exercises for a review of Chapter 11. If you have difficulty with a particular problem, review the indicated section.

1. Find the distance between the points $(-2, 2)$ and $(4, 10)$. *(Section 11.1)*

2. Find the coordinates of the midpoint of the line segment connecting $(-2, -4)$ and $(8, 10)$. *(Section 11.1)*

3. Write an equation of the perpendicular bisector of the line segment connecting the points $(5, -1)$ and $(1, 2)$. *(Section 11.1)*

4. Find a value of k so that the points $(-2, -3)$ and $(k, 2)$ are 13 units apart. *(Section 11.1)*

For Exercises 5 and 6:

a. Find an equation of the parabola with the given characteristics.

b. Graph your equation from part (a). *(Section 11.2)*

5. vertex $(0, 0)$; directrix $y = 1$ **6.** focus $(3, 2)$; vertex $(1, 2)$

Name the vertex, focus, and directrix of a parabola with the given equation. Then sketch the parabola with its focus and directrix. *(Section 11.2)*

7. $y = 2x^2$ **8.** $y = -\frac{1}{8}(x - 3)^2$

For each equation of a circle in Exercises 9 and 10, identify the center and the radius. Then graph the equation. *(Section 11.3)*

9. $(x - 1)^2 + (y + 2)^2 = 144$ **10.** $x^2 + y^2 + 6y + 9 = 2$

11. Write an equation of the circle with center $C(2, -3)$ and radius 9. *(Section 11.3)*

For Exercises 12 and 13, identify the center, the vertices, and the foci of each ellipse. Tell whether the ellipse is *horizontal* or *vertical*. Then graph the equation. *(Section 11.4)*

12. $\dfrac{(x - 3)^2}{4} + \dfrac{(y + 2)^2}{1} = 1$ **13.** $\dfrac{(x - 1)^2}{4} + \dfrac{(y + 2)^2}{16} = 1$

14. Write an equation of the ellipse with foci at $(-3, 0)$ and $(3, 0)$ and major axis of length 8. *(Section 11.4)*

Identify the center, the vertices, and the foci of each hyperbola. Tell whether the hyperbola is *horizontal* or *vertical*. Find equations of the asymptotes. Then graph the equation. *(Section 11.5)*

15. $\dfrac{(y - 1)^2}{4} - \dfrac{(x - 6)^2}{49} = 1$ **16.** $\dfrac{x^2}{4} - \dfrac{(y - 2)^2}{9} = 1$

For each equation of a conic, rewrite the equation in standard form, identify the conic, and state its important characteristics. *(Section 11.6)*

17. $2x^2 + 2y^2 - 8x + 5 = 0$

18. $y^2 - 6y + 10x - 1 = 0$

Identify and graph each equation of a degenerate conic. *(Section 11.6)*

19. $25x^2 - 4y^2 = 0$

20. $x^2 + y^2 - 4x + 2y + 5 = 0$

SPIRAL REVIEW   Chapters 1–11

1. Graph the feasible region for the system of inequalities at the right. Then find the minimum cost (in dollars) for the cost function $C = 4x + 7y$.

$$3x + y \geq 6$$
$$x + y \geq 4$$
$$x + 3y \geq 6$$
$$y \geq 0$$
$$x \geq 0$$

Describe the graph of each function and then sketch the graph.

2. $y = 3(x - 2)^2 + 1$

3. $y = -2(x + 2)^2 - 3$

Find the solutions of each quadratic equation. Use the quadratic formula.

4. $4x^2 - 3x - 2 = 0$

5. $3x^2 - 20x - 7 = 0$

Solve each system of equations.

6. $2x - y = 7$
$2x + y = 3$

7. $3x - y = 12$
$2x + 3y = 74$

Graph each function. State the domain and range.

8. $y = \sqrt{x + 1}$

9. $y = \sqrt{x - 2} + 1$

Find the slope-intercept equation of each line.

10.

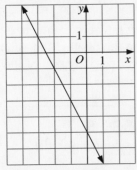

11.

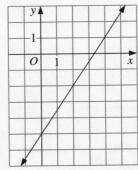

Write each function in vertex form.

12. $y = x^2 - 2x + 2$

13. $y = -2x^2 - 8x - 3$

Graph each system of inequalities.

14. $y \leq 3x - 3$
$y > x + 2$

15. $y \geq -x + 1$
$y \leq 2x + 2$

Coloring a Graph

Learn how to . . .

- represent situations with graphs

- color graphs, trying to use as few colors as possible

So you can . . .

- solve problems that require sorting items into groups

Application

Mr. Tran is setting up a schedule for the school's activity clubs. He uses a graph to determine how many different afternoons are needed to schedule the club meetings. The vertices of his graph represent the different clubs, and he draws a line (an edge) between two vertices if one or more students participate in the two clubs represented by those vertices. He then colors the vertices of the graph so that no vertices of the same color are connected by a line.

Terms to Know	*Example / Illustration*
Graph (p. 563) a set of vertices together with the edges that connect them	Yearbook · Photography · Fencing · Chess · Drawing · Astronomy · Drama
Vertices (p. 563) the points of a graph	In the graph above, the chess club and the photography club are represented by two of the seven vertices.
Edge (p. 563) a line segment that shows a connection between two vertices of a graph	In the graph above, there are 11 edges connecting the vertices.
Degree of a vertex (p. 563) the number of edges connected to a vertex	In the graph above, the vertex representing the chess club has degree 5, while the astronomy club vertex has degree 4.

Study Guide, ALGEBRA 2: EXPLORATIONS AND APPLICATIONS

UNDERSTANDING THE MAIN IDEAS

Representing situations using a graph

Graphs are used by mathematicians to give a simple visual representation of the relationships among a set of items.

Example 1

Dirk has just moved to a new town and is still learning his way around. He knows how to drive from his new house to his new school, to the town library, to the shopping center, to the nearest restaurant, and to the movie theater. He also knows the route from school to the restaurant, from school to the library, from the library to the shopping center, from the shopping center to the movie theater, and from the restaurant to the movie theater. Draw a graph that represents the routes he already knows.

Solution

Represent each of the six locations with a vertex. Connect a pair of vertices with an edge if Dirk knows the route between the locations they represent.

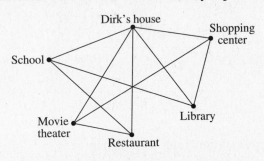

---

1. Which of the following graphs is equivalent to the one shown in Example 1?

A.

B.

C.

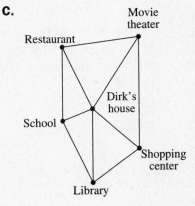

2. Dirk meets one of his friends at the library, and together they go to the restaurant for a snack. How does this change the graph in Example 1?

---

3. City Park has a beach area, concession stand, basketball court, children's playground, and a picnic area. There are walking paths from the concession stand to the beach area and to the basketball court, and from the children's playground to the picnic area and to the beach area. Draw a graph representing the park and its walking paths.

4. City Park's parking lot is connected by walking paths to each of the other points in the park. Draw a graph that includes this feature. On your graph, write the degree of each vertex.

Coloring graphs

If you need to separate items into groups that meet some criteria, a coloring algorithm may help.

Example 2

Color the graph below so that vertices that share an edge are different colors. Try to use as few colors as possible.

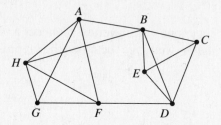

■ Solution ■

Step 1: Label each vertex of the graph with its degree.

  A, 4; *B*, 5; *C*, 3; *D*, 4; *E*, 3; *F*, 4; *G*, 3, and *H*, 4.

Step 2: Begin with the vertex with highest degree, in this case *B*. Color it red. Then color as many other vertices red as possible, so that every vertex is either red or connected to a red vertex.

Step 3: Select the vertex with highest degree that is not yet colored. Color it blue. Now color as many of the remaining vertices as possible blue. Remember, if the vertices are connected, they cannot be the same color.

(Solution continues on the next page.)

■ Solution ■ *(continued)*

Step 4: Continue in this way with additional colors until all the vertices have been colored. The figure below indicates the colors for each of the vertices.

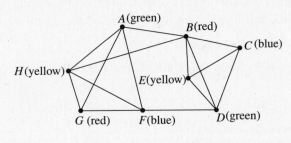

Color each graph so that vertices that share an edge are different colors. Try to use as few colors as possible.

5.

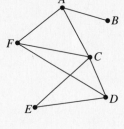

6.

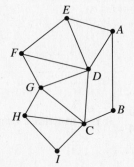

7.

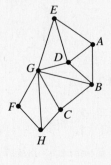

A map of nine midwestern states is shown at the right.

8. Draw a graph whose vertices represent the nine states and where an edge connects two vertices if they share a common border.

9. Color your graph using as few colors as possible.

10. Trace the map, and use your graph as a guide for coloring each state.

Oneida has a whole family of pet mice who are overcrowded in their cage. They do not get along under such crowded conditions. For convenience, she has named them A, B, C, D, E, F, G, and H. She has noticed that A does not get along with C, E, F, and G; B does not get along with C, D, and F; and C does not get along with G and H.

11. Draw a graph whose vertices represent the eight mice, with edges connecting two vertices if the mice they represent do not get along.

12. Color your graph using as few colors as possible.

13. What does your colored graph indicate about how Oneida should separate her mice into new cages?

....
Spiral Review

For each equation of a conic, rewrite the equation in standard form, identify the conic, and state its important characteristics. *(Section 11.6)*

14. $3x^2 + 3y^2 + 6x = 24$

15. $x^2 - 9y^2 - 8x - 36y = 29$

16. $x = 0.5y^2 - y - 2.5$

17. $4x^2 + y^2 + 24x + 32 = 0$

Solve. *(Section 8.3)*

18. $\sqrt{x} = 5$

19. $\sqrt{x + 2} = 3$

20. $\sqrt{6 - 2x} = 4$

Find each product for $A = \begin{bmatrix} 2 & -1 \\ -3 & 1 \end{bmatrix}$ **and** $B = \begin{bmatrix} 0 & 2 & 0 \\ -1 & -3 & 4 \end{bmatrix}$. **If the matrices cannot be multiplied, state that the product is** *undefined*. *(Section 1.4)*

21. AB

22. BA

23. A^2

Directed Graphs and Matrices

Learn how to . . .

- represent situations with directed graphs

- represent directed graphs with matrices

So you can . . .

- analyze situations involving direction

Application

The Heartbreakers soccer team has a telephone tree to use if a game must be cancelled or postponed. The coach calls Alyssa, Beth, and Carmen. Alyssa calls Danielle, Danielle calls Erin, and Erin calls Fayola. Beth calls Gloria, Gloria calls Holly, and Holly calls Ida. Carmen calls Juana, Juana calls Karin, and Karin calls Lani. If a girl cannot reach the teammate she is supposed to call, she calls the next name on the list instead. The girl at the end of each branch of the tree calls the coach so the coach knows that everyone has been contacted. This telephone tree can be represented by a directed graph.

Terms to Know	**Example / Illustration**
Directed graph (p. 569) a graph in which the edges are arrows	The graph below represents the telephone tree discussed in the Application. 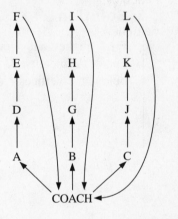

UNDERSTANDING THE MAIN IDEAS

Directed graphs can be used to represent many types of relationships, for example food webs, one-way streets or the prerequisites for different math courses.

In college, there are many courses that students cannot enroll in until they have taken another course or courses first. A course that must be taken first is called a *prerequisite* for the subsequent course. A college mathematics department lists the following course prerequisites.

Course	Prerequisite(s)
Math 121 (Calculus I)	Math 105 (College Algebra)
Math 122 (Calculus II)	Math 121
Math 123 (Calculus III)	Math 122
Math 151 (Discrete Math)	Math 105
Math 201 (Linear Math)	Math 122 and Math 151
Math 221 (Calculus IV)	Math 123
Math 301 (Abstract Algebra I)	Math 201
Math 302 (Abstract Algebra II)	Math 301
Math 321 (Real Analysis I)	Math 221
Math 322 (Real Analysis II)	Math 321

Represent these course relationships using a directed graph.

■ Solution ■

Each mathematics course is represented by a vertex, and then an arrow is drawn *from* course A *to* course B if course A is a prerequisite for course B. In the directed graph below, the vertices are labeled with the course numbers.

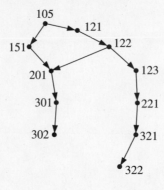

1. The following courses and their prerequisites are listed in the college's summer course catalog. Add them to the directed graph shown in Example 1.

Math 331 (Mathematical Statistics I): Math 221 and Math 151
Math 332 (Mathematical Statistics II): Math 331

2. Which mathematics courses (including Math 331 and Math 332 listed in Exercise 1) depend directly on Math 122? Which courses depend on Math 122 indirectly?

3. **Open-ended** Think of a situation that can be modeled by a directed graph. Draw a graph of the situation, explaining what the vertices and edges represent.

4. The children at a birthday party are playing a game of tag. Freddie was "it" at the start of the game. Freddie tagged Randhir, Randhir tagged Maria, Maria tagged Carolyn, Carolyn tagged Winton, Winton tagged Randhir, Randhir tagged Freddie, and Freddie had just tagged Maria when everyone was called in to have cake and ice cream. Draw a directed graph where each child is a vertex, and an arrow is drawn from vertex *A* to vertex *B* if child *A* tagged child *B*.

Matrices of directed graphs

Directed graphs can be represented by matrices whose elements are 1s and 0s, based on the relationships between each pair of vertices. After representing a directed graph by a matrix, you can use technology to find direct and indirect relationships between vertices.

Example 2

Construct a matrix to represent the directed graph in Example 1.

■ **Solution** ■

The rows and columns of the matrix correspond to the vertices of the graph. An element of the matrix is a 1 if the vertex represented by that row has an arrow going to the vertex represented by that column. All the other elements of the matrix are 0.

	105	121	122	123	151	201	221	301	302	321	322
105	0	1	0	0	1	0	0	0	0	0	0
121	0	0	1	0	0	0	0	0	0	0	0
122	0	0	0	1	0	1	0	0	0	0	0
123	0	0	0	0	0	0	1	0	0	0	0
151	0	0	0	0	0	1	0	0	0	0	0
201	0	0	0	0	0	0	0	1	0	0	0
221	0	0	0	0	0	0	0	0	0	1	0
301	0	0	0	0	0	0	0	0	1	0	0
302	0	0	0	0	0	0	0	0	0	0	0
321	0	0	0	0	0	0	0	0	0	0	1
322	0	0	0	0	0	0	0	0	0	0	0

Write a matrix representing each directed graph. Let an element be 1 if the row vertex has an edge directed to the column vertex. Otherwise, let the element be 0.

5.

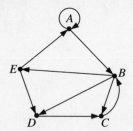

6.

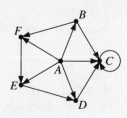

7.

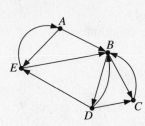

Katie found out that her math teacher's birthday is next Friday. She told Zachary, another student in her class. He spread the news to Michelle, Albert, and Juanita. Michelle told Julie and Ferris, Albert told Cochita and Anne, and Juanita told Silas, Carl, and Arlene. Use this information for Exercises 8 and 9.

8. Model this situation with a directed graph.

9. Write a matrix representing your directed graph for Exercise 8.

10. Mathematics Journal What are the major features of a graph as it is used in discrete mathematics? Name some situations that can be modeled by graphs and directed graphs, and explain how the graphs are used to answer questions or solve problems.

· · · · · · · · · · · · · · · · · · ·
Spiral Review

11. Tonya is planning to play charades at her next party. Some of her guests do not get along well, and she wants to decide on teams ahead of time. Arla does not get along with Stu or Bill. Bill does not get along with Carla, Frank, or Pat. Stu is not speaking to Pat, and does not get along with Jean or Don either. Tonya and her best friend, Gayle, get along with everyone. What is the best way to divide the group up into teams? Assume that the teams do not need to have an equal number of members. *(Section 12.1)*

Find the sum of each series, if the sum exists. If a series does not have a sum, explain why not. *(Section 10.5)*

12. $0.1 + 0.11 + 0.111 + \cdots$

13. $\dfrac{1}{4} + \dfrac{3}{8} + \dfrac{9}{16} + \cdots + \dfrac{243}{128}$

14. $9 - 3 + 1 - \dfrac{1}{3} + \cdots$

15. $4 + 7 + 10 + 13 + \cdots + 82$

Permutations

Application

An ice cream shop has 16 different flavors of ice cream and frozen yogurt available. Suppose you want a double cone. You choose one flavor for the bottom scoop and a second flavor for the top scoop. There are 16×15, or 240 different 2-scoop combinations that are possible.

Learn how to . . .

- use the multiplication counting principle
- find the number of permutations of the elements of a set

So you can . . .

- count the possibilities in situations

Terms to Know	Example / Illustration
Permutation (p. 579) an arrangement of the elements of a set	The 16 flavors of ice cream at the shop can be arranged in $16! \approx 2.09 \times 10^{13}$ different ways in the display case.

UNDERSTANDING THE MAIN IDEAS

The multiplication counting principle

The multiplication counting principle is a rule for determining how many different arrangements are possible when a series of choices are made. If the first choice is among m different alternatives and the second choice is among n different alternatives, then there are a total of $m \times n$ possibilities. This rule can be extended to a series of more than two choices as well.

Example 1

Kimiko is packing her lunch for tomorrow. She can choose peanut butter, ham or tuna salad for a sandwich, have carrot sticks or cole slaw for a vegetable, and pick an orange, apple, or banana for a fruit. How many different lunches containing one sandwich, one vegetable, and one fruit are possible?

There is a series of three choices for Kimiko to make. Apply the multiplication principle as follows.

$$\begin{pmatrix} \text{ways to choose} \\ \text{a sandwich} \end{pmatrix} \times \begin{pmatrix} \text{ways to choose} \\ \text{a vegetable} \end{pmatrix} \times \begin{pmatrix} \text{ways to choose} \\ \text{a fruit} \end{pmatrix}$$

$$3 \quad \times \quad 2 \quad \times \quad 3 \quad = 18$$

Kimiko can choose from 18 different lunches.

---

1. Suppose Kimiko finds a carton of yogurt in the refrigerator as well, and can choose between it and one of the sandwiches. How many different lunches are possible now?

2. The school cafeteria offers two different entrees for lunch, along with a choice of salad or fruit, and regular or low fat milk. How many ways can a lunch be selected from this menu?

3. Maxim is treating his parents to an evening out to celebrate their anniversary. They can select one of four different moderately priced restaurants, and then pick one of six movies showing at a nearby multiplex theater. How many different ways can they select a restaurant and a movie?

In one state, license plates have three letters followed by three digits.

4. How many license plates can be made if the letters and digits can be repeated?

5. How many license plates can be made if no digits or letters can be repeated? What percent of your answer to Exercise 4 is this?

Permutations

A *permutation* is an arrangement of the elements of a set. If all n of the elements of a set are to be arranged, there are $n!$ different arrangements. (*Note*: 0! is defined to be equal to 1.) However, if some of the elements of a set are repeated, some of the arrangements will look the same. Therefore, if among the n elements in a set, one element is repeated q_1 times, another is repeated q_2 times, ..., and the last element is repeated q_k times, then there are $\dfrac{n!}{q_1! q_2! ... q_k!}$ distinguishable arrangements of the elements in the set.

The number of possible arrangements of the elements of a set is different if not all the elements will be selected. The symbol $_nP_r$ is used to represent the number of permutations of r objects taken from a set of n objects, all of which are different. The formula for $_nP_r$ is $\dfrac{n!}{(n-r)!}$.

Example 2

How many distinguishable permutations are there of the letters in each word?

a. ZOT   **b.** ZOO   **c.** ZOOLOGICAL

■ Solution ■

a. Since the three letters in ZOT are all different, there are 3!, or 6 distinguishable permutations: ZOT, ZTO, OZT, OTZ, TZO, and TOZ.

b. Since the letters in ZOO are not all different, there are fewer distinguishable arrangements of them than for the letters in ZOT. Since the O is repeated 2 times, there are $\frac{3!}{2!} = \frac{6}{2}$, or 3 distinguishable permutations.

c. The word ZOOLOGICAL contains 10 letters, 3 of which are Os and 2 of which are Ls. So, there are $\frac{10!}{3!2!}$, or 302,400 distinguishable permutations.

Simplify.

6. 5!   **7.** 7!   **8.** $_8P_5$ **9.** $_6P_4$

For Exercises 10–12, find the number of distinguishable permutations of each group of letters.

10. SPRING   **11.** SUMMER   **12.** SUCCESS

13. A basketball team has 12 players. In how many ways can they line up to enter the gym at the beginning of the game? If the captain is always first and the co-captain is always second in line, in how many ways can the remaining players line up?

14. A school librarian sets up a special display of books each week. This week, he plans to feature books about spring sports. He has a collection of 32 books to choose from, but space to display only 5 of them. In how many ways can he arrange 5 of the 32 books in the display area?

15. An English teacher has reserved parts of 3 class periods for his students to give oral book reports. He will select 9 of the 27 students to report on the first day. In how many ways can he select 9 students to give their reports?

---

16. A rather complicated computer program consists of the main program A and a number of subroutines, designated B, C, D, E, and F. Program A can call subroutines B, C, and D directly. Subroutine B can call itself as well as subroutines C and D. Subroutines C and D can each call subroutines E and F, and subroutines E and F can call themselves. *(Section 12.2)*

a. Draw a directed graph representing this situation.

b. Write a matrix that represents the graph. Let an element be 1 if the row vertex has an edge directed to the column vertex. Otherwise, let the element be 0.

Find the asymptotes of the graph of each function. Tell how the graph is related to a hyperbola with equation of the form $y = \dfrac{a}{x}$. *(Section 9.7)*

17. $f(x) = \dfrac{3}{x-2}$

18. $h(x) = \dfrac{x-1}{2x+2}$

19. $g(x) = \dfrac{3x}{x-2}$

Combinations

Learn how to . . .

- find the number of
  combinations of the
  elements of a set

So you can . . .

- count choices in
  which order is not
  important

Application

There are openings for 4 students on the yearbook committee and 12 qualified students have applied. The order in which the students are selected is not important. There are $_{12}C_4 = 495$ different groups of four applicants that can be selected.

Terms to Know	**Example / Illustration**
Combination (p. 585) a selection of r items from a set of n items when position is not important	If you must read 3 books from a suggested reading list of 10 books, there are $_{10}C_3 = 120$ different groups of 3 books you may choose.

UNDERSTANDING THE MAIN IDEAS

The formula for finding combinations

The number of combinations of r items that can be selected from a set of n items

is $_nC_r = \dfrac{n!}{(n-r)!\,r!}$.

Example 1

Show that there are 120 possible groups of 3 books that can be selected from a reading list of 10 books.

◾ Solution◾

The order in which the 3 books are chosen makes no difference. It is the same group of 3 books no matter what order you read them in. This is a combination of 3 items selected from a set of 10 items. Use the formula $_nC_r = \dfrac{n!}{(n-r)!r!}$ with $n = 10$ and $r = 3$.

$$\frac{10!}{3!7!} = \frac{10 \cdot 9 \cdot 8 \cdot \cancel{7!}}{3!\cancel{7!}}$$

$$= \frac{10 \cdot 9 \cdot 8}{3!}$$

$$= \frac{10 \cdot 9 \cdot 8}{6}$$

$$= 120$$

So, there are 120 ways to select a group of 3 books to read from the list of 10 books.

For Exercises 1–5, simplify.

1. $_4C_2$ 2. $_5C_3$ 3. $_5P_3$ 4. $_{10}C_5$ 5. $_{10}C_9$

6. Suppose every student in a literature class is required to read a certain one of the 10 books on a list, and then may chose any two of the other nine books on the list. How many combinations of books may be selected?

7. Samara wants to have a build-your-own-sundae party to celebrate her birthday. She wants to buy four different flavors of ice cream. The store has 15 flavors available. In how many ways can she pick four flavors to serve at her party?

8. Robert has room to plant two varieties of tomatoes in the garden in his back yard. He is considering five different varieties at the local nursery. In how many ways can he choose two varieties from the five available?

Counting more than one type of choice

Some problems require the use of the multiplication counting principle or addition, as well as a permutation or a combination. When first considering a problem, determine whether the different choices are made in a certain order or the different choices just form one group from which a single selection is made.

Example 2

A restaurant offers several types of soup each day. Each day's soup selections are chosen from the vegetarian soups cream of broccoli, split pea, minestrone, and vegetarian chili, as well as from the non-vegetarian soups clam chowder, vegetable beef, chicken noodle, beef stew, and regular chili.

a. How many combinations of four soups are possible from the list of soups above?

b. How many combinations are possible if the restaurant offers two vegetarian soups and two non-vegetarian soups?

■ Solution ■

a. There are nine soups from which to make the selections, and order is not important. The number of combinations of four soups chosen from a list of nine soups is $_9C_4 = \dfrac{9!}{5!4!}$, or 126.

There are 126 combinations of four soups that can be chosen from the list.

b. In this situation, the restaurant selects two of the four vegetarian soups and two of the five non-vegetarian soups. There are $_4C_2 = 6$ ways to choose the vegetarian soups, and $_5C_2 = 10$ ways to select the non-vegetarian soups.

Therefore, there are 6×10, or 60 combinations of two vegetarian and two non-vegetarian soups that can be chosen.

---

9. On Fridays, the restaurant always serves clam chowder. In how many ways can the restaurant offer four soups, with one more being non-vegetarian and the other two being vegetarian soups?

10. There are three centers, five forwards, and six guards among the players on the high school basketball team. At any given time, one center, two forwards, and two guards are playing in the game. In how many ways can a group of five players be selected?

11. Naomi enjoys entering sweepstakes and other contests. She knows of five different contests going on right now, but she only has three postage stamps. In how many ways can Naomi select three of the five contests to enter?

12. Steve is interested in an offer from a software company, where he can get one of five games free if he purchases two others at the regular price. There are 15 games to choose among for the two he must buy. In how many ways can he select his three games?

13. The judges of an art show must select three entries from among the 31 artworks to be entered in the Best of Show category. In the past, four entries have sometimes been selected to enter the final round if the judges felt that all were deserving of the honor. In how many total ways can the judges select the Best of Show entries if either three or four artworks are selected?

Etenia has purchased one ticket for a raffle in which a total of 200 tickets were sold. Five prizes are to be awarded.

14. In how many ways can five winning tickets be selected from the 200 tickets sold?

15. How many of the possible groups of five winning tickets include Etenia's ticket? (*Hint*: If Etenia's ticket is one of the five selected, then the other four winning tickets are drawn from the remaining 199 tickets.)

· ·
Spiral Review

Simplify each expression. (*Section 12.3*)

16. $_4P_1$

17. $_5P_3$

18. $_{10}P_4$

19. $_{20}P_4$

20. $_{12}P_5$

21. $_8P_1$

Multiply. (*Section 8.2*)

22. $(3x - 7)(2x + 10)$

23. $(x - 4)^3$

24. $(t - 3)(t^2 + 2t + 5)$

25. $(m + 1)(m^2 - 1)$

26. $3y(y - 6)(4 - y)$

27. $(5y - 6)^2$

Find an equation for each cubic function whose graph is shown. (*Section 9.4*)

28.

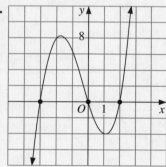

29.

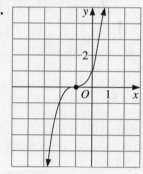

Pascal's Triangle

Application

Each day Rosa takes her dog for a walk from her house on the corner of 1st Avenue and Washington Street to the park, whose main entrance is on the corner of 5th Avenue and Monroe Street. To get to the main entrance, Rosa and her dog must walk 4 blocks south and 4 block east, for a total of 8 blocks. There are $_8C_4$, or 70 possible routes they can take. The number of routes to various points on this street grid is one model of Pascal's triangle.

Terms to Know	Example / Illustration
Pascal's triangle (p. 593) a triangular display of numbers with the property that term r in row n, where r and n start at 0, is given by $_nC_r$	The first five rows of Pascal's triangle are shown below. 1 1 1 1 2 1 1 3 3 1 1 4 6 4 1

UNDERSTANDING THE MAIN IDEAS

Pascal's triangle and combinations

The nth row of Pascal's triangle is made up of the numbers $_nC_r$, for $r = 0$ to n. When writing out Pascal's triangle, you can generate the values by addition using the following relationship. The ith number in row n is the sum of the $(i - 1)$st and ith numbers in row $n - 1$. In other words, except for the first and last elements (which are always 1) in any row, each element in the triangle is the sum of the two numbers closest to it in the row above.

Example 1

Construct Pascal's triangle up to row 6.

■ Solution ■

Remember, every row of Pascal's triangle starts and ends with a 1. The first row of the triangle is called row 0. Apply either the addition rule given above or use $_nC_r$ for $n = 0$ to 6 and $r = 0$ to n.

Row 0							1						
Row 1						1		1					
Row 2					1		2		1				
Row 3				1		3		3		1			
Row 4			1		4		6		4		1		
Row 5		1		5		10		10		5		1	
Row 6	1		6		15		20		15		6		1

1. Determine rows 7 and 8 of Pascal's triangle.

2. How many elements are in row n of Pascal's triangle?

3. Which row of Pascal's triangle begins 1, 12, 66, ...? Explain how you know.

4. Which rows have two consecutive numbers in the middle that are the same?

5. Refer to the map and discussion in the Application. Suppose there is another park entrance at the corner of 8th Avenue and Monroe Street. How many different routes can Rosa and her dog walk in order to reach this entrance?

Pascal's triangle and the binomial theorem

To find the expanded form of the power of a binomial, use the binomial theorem.

For any positive integer n,

$$(a + b)^n = {_nC_0}a^n + {_nC_1}a^{n-1}b + {_nC_2}a^{n-2}b^2 + \cdots + {_nC_{n-1}}ab^{n-1} + {_nC_n}b^n$$

Notice that the coefficients of the terms in the expansion are the elements of row n in Pascal's triangle. For example, to write the expanded form of $(x + y)^4$, the coefficients of the terms of the expansion are the elements of row 4.

Example 2

Find the expanded form of $(x + 2)^4$.

■ Solution ■

Use the binomial theorem with $a = x$, $b = 2$, and $n = 4$.

$$(x + 2)^4 = {_4C_0}x^4 + {_4C_1}x^3(2) + {_4C_2}x^2(2)^2 + {_4C_3}x(2)^3 + {_4C_4}(2)^4$$
$$= (1 \cdot x^4) + (4 \cdot x^3 \cdot 2) + (6 \cdot x^2 \cdot 4) + (4 \cdot x \cdot 8) + (1 \cdot 16)$$
$$= x^4 + 8x^3 + 24x^2 + 32x + 16$$

Give the values of *a*, *b*, and *n* that you would substitute in the binomial theorem to expand each power of a binomial.

6. $(2x - 1)^7$ **7.** $(s + 3t^2)^8$ **8.** $(x - yz)^5$

For Exercises 9–12, use the binomial theorem to expand each power of a binomial.

9. $(x + y)^4$ **10.** $(a + 3)^5$ **11.** $(x - 5y)^3$ **12.** $(a - 2b)^6$

13. You can use the binomial theorem to estimate the value of a number raised to a power, such as $(0.99)^{10}$.

  a. Since $0.99 = 1 - 0.01$, then $(0.99)^{10} = (1 - 0.01)^{10}$. Use the binomial theorem to write this power in expanded form.

  b. What happens to the value of the terms in this expansion as the power of 0.01 increases?

  c. Use your expansion from part (a) to estimate the value of $(0.99)^{10}$ to the nearest thousandth.

14. Use the method of Exercise 13 to estimate $(1.02)^7$ to the nearest hundredth.

15. **Mathematics Journal** Explain how $_nP_r$ and $_nC_r$ are different and how they are the same. Use an example to illustrate your answer.

....

Spiral Review

For Exercises 16–19, find the value of each expression. *(Section 12.4)*

16. $_5C_2$ **17.** $_7C_4$ **18.** $_{10}C_4$ **19.** $_{20}C_{10}$

20. A cylinder is 5 cm high and 8 cm in diameter. Find the surface area of the cylinder. *(Toolbox, page 802)*

Write a point-slope equation of the line passing through the given point and having the given slope. *(Section 2.3)*

21. point: (2, 0); slope = 0.5 **22.** point: (–1, 3); slope = 3

23. point: (4, 5); slope = 0 **24.** point: (–2, –1); slope = $\frac{3}{4}$

Complete these exercises for a review of Chapter 12. If you have difficulty with a particular problem, review the indicated section.

Elena needs to contact each member of her drama club about an upcoming rehearsal. The club members are scattered around several communities, so that calling some members is long distance. Elena can call Mark, Chandra, and Kerri directly. Mark can call Becky, Leanne, and Alvin, and both Becky and Alvin can call Tricia and Saul. Kerri can call Mina, Charlene, and Louis. Use this information for Exercises 1 and 2. *(Section 12.1)*

1. Make a graph that models this situation.

2. What is the greatest number of calls it takes to reach any member of the club? Who is the most "distant" from Elena in this sense?

3. **Open-ended** Draw a map showing several different areas. Represent your map by a graph and then determine a coloring scheme for the map. Then color your map using as few colors as possible. *(Section 12.1)*

4. **Writing** Explain how a board game might be represented by a directed graph. What would the vertices and arrows represent? *(Section 12.2)*

5. Write a matrix representing the directed graph at the right. *(Section 12.2)*

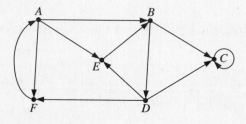

6. Jennifer is packing her suitcase for a trip. She decides to bring along one of three new magazines she has not read yet, one of four books she just got from the library, and one of her two favorite travel games. In how many different ways can she select these items? *(Section 12.3)*

7. The organizer of a 4th of July parade must decide the order in which the floats, bands, and other units will be placed. She decides to lead off with the high school marching band, and then have three of the 15 floats follow before the next marching band. In how many ways can she arrange three of the 15 floats to follow the high school band? *(Section 12.3)*

8. Hamadi is making his mother a necklace for Mother's Day. He wants to create a pattern with 3 pink, 2 green, and 2 blue beads, and then repeat the pattern over and over to make a necklace of the desired length. How many different arrangements can he make with his basic set of 7 beads? *(Section 12.3)*

9. As advisor to the senior class, Ms. Pyle must select three students from a group of seven applicants to speak at the graduation ceremony. In how many ways can she choose the three speakers? *(Section 12.4)*

Ralph is allowed one hour of television, video game time, or computer game time each day. There is a one-hour show he sometimes watches on television tonight, four computer games he is currently trying to master, and six video games he can borrow from a friend. Use this information for Exercises 10 and 11. *(Section 12.4)*

10. Ralph knows he can play two video games or two computer games in one hour. If he can choose the television show or any two of the games, in how many different ways can he spend his one hour today?

11. If Ralph can either play computer games or play video games, but not both, and he has decided not to watch the television show, in how many ways can he spend his one hour?

12. Which row of Pascal's triangle begins with the numbers 1, 8, 28? What are the next two elements of this row? *(Section 12.5)*

13. Use the binomial theorem to expand $(x - 3y)^5$. *(Section 12.5)*

SPIRAL REVIEW   Chapters 1–12

For Exercises 1–3, find the inverse of each function, if it exists. If there is no inverse, tell why. If there is an inverse, give its domain and range.

1. $y = \frac{2}{3}x - 5$ **2.** $y = \log_2 x$ **3.** $y = e^x$

4. Find an equation for the polynomial function that has a double root at $x = -4$ and whose graph passes through the origin.

Use the sequence 1, 3, 6, 10, 15, ...

5. What are the next four terms of the sequence? Give a recursive formula for the nth term.

6. What is the sum of the first 10 terms of the related series?

7. Write out several rows of Pascal's triangle and then explain how this group of numbers is related to this sequence.

For Exercises 8–10, simplify.

8. $(3 + 4i)(2 - i)$ **9.** $(2 - i)(2 + i)$ **10.** i^4

11. Find all real and imaginary roots of the polynomial function $f(x) = x^4 - 2x^3 - 6x - 9$.

12. Find an equation of the line perpendicular to the line $y = 2x - 5$ and passing through the point $(1, 4)$.

13. Find the distance between the points $(-2, 5)$ and $(2, -5)$.

The assessed values of houses located on one city block are $119,000, $129,500, $124,400, $134,500, $149,100, $139,000, $127,500, $135,900, $144,000, and $127,300.

14. Find the mean, the median, and the standard deviation of the data set.

15. Construct a box plot of the data and find the interquartile range.

Exploring Probability

Learn how to . . .

- calculate experimental, theoretical, and geometric probabilities

So you can . . .

- predict the outcome of different events

Application

Carlos has had 22 hits in his first 91 at-bats this baseball season. The experimental probability of a hit on his next at-bat is $\frac{22}{91} \approx 0.242$, or about 24%.

Terms to Know

Example / Illustration

Probability (p. 611) the fraction of the time that an event is expected to happen (The probability of event A is denoted $P(A)$.)	When rolling a die, $P(4) = \frac{1}{6}$ and $P(\text{an odd number}) = \frac{1}{2}$.
Experimental probability (p. 612) the probability of an event found by performing an experiment consisting of a certain number of trials $P(A) = \dfrac{\text{number of trials where } A \text{ happens}}{\text{total number of trials}}$	In the Application above, the experimental probability of Carlos getting a hit on any at-bat is $\frac{22}{91} \approx 0.242$.
Theoretical probability (p. 613) the probability of an event based on the number of ways an event can occur and the number of equally-likely outcomes $P(A) = \dfrac{\text{number of ways } A \text{ can occur}}{\text{total number of possible outcomes}}$	If there are 22 red balls among the 91 balls in an urn, the probability of drawing a red one is $\frac{22}{91} \approx 0.242$.
Geometric probability (p. 614) the probability of an event that is found by comparing the areas of geometric figures	The probability of the pointer landing on red when spinning the spinner below is $\frac{1}{4}$.

UNDERSTANDING THE MAIN IDEAS

Experimental and theoretical probability

You find an experimental probability based on the results of several observations or trials. Theoretical probability is calculated by comparing the number of possible outcomes of an event to the total number of possible outcomes.

Example 1

Joan rolled a die 30 times and obtained the following results.

Outcome	1	2	3	4	5	6
Frequency	5	7	5	5	4	4

a. What is the experimental probability of rolling a 5? a number less than 3? an odd number?

b. What is the theoretical probability of rolling a 5? a number less than 3? an odd number?

■ Solution ■

a. Joan obtained four 5's in 30 rolls, so $P(5) = \frac{4}{30} \approx 0.133$.

A 1 occurred on five of the rolls and a 2 occurred on seven of the rolls, so $P(\text{a number less than 3}) = \frac{12}{30} \approx 0.4$.

A 1 occurred five times, a 3 occurred five times, and a 5 occurred four times, for a total of 14 odd numbers in 30 trials, so $P(\text{odd number}) = \frac{14}{30} \approx 0.467$.

b. Theoretically, the outcomes 1, 2, 3, 4, 5, and 6 are all equally likely, so each has a probability of $\frac{1}{6}$. Therefore, $P(5) = \frac{1}{6}$; $P(\text{number less than 3}) = P(1 \text{ or } 2) = \frac{2}{6}$, or $\frac{1}{3}$; and $P(\text{odd number}) = P(1, 3, \text{ or } 5) = \frac{3}{6}$, or $\frac{1}{2}$.

In 20 flips of a fair coin, the coin landed heads 13 times. Use this information for Exercises 1 and 2.

1. What is the experimental probability of the coin landing heads on a flip of the coin? of the coin landing tails?

2. What is the theoretical probability of the coin landing heads on a flip of the coin? of the coin landing tails?

3. Mariko had 28 hits in her first 80 at-bats this softball season. Find the probability that she will get a hit on her next at-bat.

4. Roberto puts his spare change in a glass jar each evening before going to bed. So far, the jar contains 3 quarters, 7 dimes, 5 nickels, and 27 pennies. What is the probability that the next coin Roberto puts in the jar will be a nickel? a penny?

5. Suppose Roberto reaches into his jar and pulls out a coin at random. Assuming that each of the coins is equally likely to be chosen, what is the probability that he will pull out a nickel? a penny?

6. What type of probability is used to answer Exercise 4? What type of probability is used to answer Exercise 5?

Geometric probability

Sometimes probabilities are found by comparing the areas of geometric figures.

Example 2

The apple tree in Jamal's backyard overhangs his patio. When an apple drops off the tree, it may hit the cement patio and be ruined. Referring to the figure at the right, find the probability that a falling apple will hit the patio. Assume the apple falls straight down.

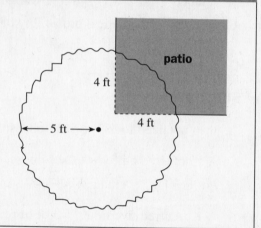

■ Solution ■

First, you need to find the approximate area of the patio that lies directly below the apple tree, where an apple might fall. From the figure, the part of the patio below the tree approximates a sector of a circle that has radius 4 ft. The area of this sector appears to be approximately one-fourth of the area of an entire circle of radius 4 ft.

$$\text{area of patio below tree} \approx \frac{1}{4}\pi(4)^2$$

$$= \frac{1}{4}\pi(16)$$

$$\approx 12.57 \text{ ft}^2$$

The total area below the tree is approximately that of a circle of radius 5 ft.

(Solution continues on next page.)

Thus, P(falling apple hits patio) $= \dfrac{\text{area of patio below tree}}{\text{total area below tree}}$

$$\approx \dfrac{12.57}{\pi(5^2)}$$

$$\approx \dfrac{12.57}{78.54}$$

$$\approx 0.160$$

The probability that an apple falling from the tree will land on the patio is about 0.160.

7. If 66 apples randomly fall from Jamal's tree before the crop is harvested, about how many will be ruined by hitting the patio?

8. The spinner at the right is spun once. What is the probability that the pointer stops on red? on blue?

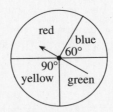

The targets shown are squares. Suppose a randomly thrown dart hits each target. Find the probability that the dart hits the target's shaded region. (The shaded regions are formed by squares and circles.)

9.

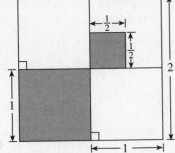

10.

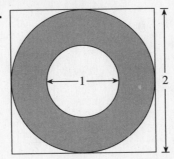

·····················
Spiral Review
·····················

Use the binomial theorem to expand each power of a binomial. *(Section 12.5)*

11. $(x + 2)^3$ **12.** $(a - 2b)^4$ **13.** $(3 + t^2)^5$

Find all real and imaginary zeros of each polynomial function. Identify any double or triple zeros. *(Sections 9.4 and 9.5)*

14. $f(x) = x^3 + x^2 - 16x + 20$ **15.** $f(x) = x^4 - 3x^3 - x^2 - 9x - 12$

Working with Multiple Events

Learn how to . . .

- find probabilities involving independent, mutually exclusive, and complementary events

So you can . . .

- solve problems involving the probability of multiple events

Application

Elena just planted two pumpkin seeds in her vegetable garden. If the seeds have a 90% germination rate, the probability that both seeds will grow is $(0.9)(0.9) = 0.81$. This is an example of independent events.

Terms to Know

Example / Illustration

Independent events (p. 619) two events A and B for which the occurrence of A does not affect whether B happens, and the occurrence of B does not affect the occurrence of A (rule: $P(A \text{ and } B) = P(A) \cdot P(B)$)	If a coin is flipped twice in a row, the flips are independent. So, $P(H_1 \text{ and } H_2)$ $= P(H_1) \cdot P(H_2)$ $= \dfrac{1}{2} \cdot \dfrac{1}{2}$ $= \dfrac{1}{4}$
Mutually exclusive events (p. 619) events that cannot happen at the same time (rule: $P(A \text{ or } B) = P(A) + P(B)$)	When a die is rolled, the events "rolling a 4" and "rolling a 6" are mutually exclusive, since they cannot happen at the same time.
Complementary events (p. 619) two events that are mutually exclusive and one of which must happen (rule: $P(B) = 1 - P(B)$)	When a die is rolled, the complement of the event "rolling an odd number" is the event "rolling an even number."

UNDERSTANDING THE MAIN IDEAS

Independent and mutually exclusive events

If two events are independent, then the probability that *both* events occur is the *product* of the probabilities that each event occurs: $P(A \text{ and } B) = P(A) \cdot P(B)$.

If two events are mutually exclusive, then the probability that *one or the other* event occurs is the *sum* of the probabilities that each event occurs: $P(A \text{ or } B) = P(A) + P(B)$. (*Note:* If events A and B are mutually exclusive, then they cannot both occur and $P(A \text{ and } B) = 0$.)

Example 1

An urn contains 5 red balls, 3 black balls, and 2 white balls. An experiment consists of drawing one ball from the urn, noting its color, returning it and remixing the balls, and then drawing a second ball. Let A be the event that the first ball is red, let B be the event that the second ball is black, and let C be the event that the second ball is white. Find each indicated probability.

a. $P(A)$ **b.** $P(B)$ **c.** $P(C)$ **d.** $P(A \text{ and } B)$

e. $P(A \text{ and } C)$ **f.** $P(B \text{ and } C)$ **g.** $P(B \text{ or } C)$

■ Solution ■

a. $P(A) = \dfrac{\text{number of red balls}}{\text{total number of balls}} = \dfrac{5}{10} = 0.5$

b. $P(B) = \dfrac{\text{number of black balls}}{\text{total number of balls}} = \dfrac{3}{10} = 0.3$

c. $P(C) = \dfrac{\text{number of white balls}}{\text{total number of balls}} = \dfrac{2}{10} = 0.2$

d. Since A and B are independent events, $P(A \text{ and } B) = P(A) \cdot P(B)$. Using the results of parts (a) and (b), $P(A \text{ and } B) = (0.5)(0.3) = 0.15$.

e. Since A and C are also independent events, $P(A \text{ and } C) = P(A) \cdot P(C)$. Using the results of parts (a) and (c), $P(A \text{ and } C) = (0.5)(0.2) = 0.1$.

f. Since the second ball drawn cannot be both black and white, $P(B \text{ and } C) = 0$. (Events B and C are mutually exclusive.)

g. Since events B and C are mutually exclusive, $P(B \text{ or } C) = P(B) + P(C)$. Using the results of parts (b) and (c), $P(B \text{ or } C) = 0.3 + 0.2 = 0.5$.

For Exercises 1 and 2, a coin is flipped three times.

1. Find P(three heads) and P(three tails) if the coin is a fair coin.

2. Find P(three heads) and P(three tails) if the coin is biased so that P(heads) on any particular toss is 0.7.

3. An ordinary die is rolled twice. Let A be the event that the first roll is a 5 or 6 and let B be the event that the second roll is a 1, 2, or 3. Find $P(A)$, $P(B)$, and $P(A \text{ and } B)$.

4. In her rush to get to her job, Rosa forgets to take her umbrella about 30% of the time. If the weatherman says there is a 70% chance of rain, what is the probability that it rains and Rosa left her umbrella at home?

5. A raffle is held in which two lucky people will win a dinner for two at a local restaurant, and five other people will win a free ticket for a movie at the town cinema. If 500 raffle tickets are sold, what is the probability of winning a dinner for two? winning a cinema ticket? winning any prize?

For Exercises 6 and 7, an experiment consists of rolling a die three times and recording the result of each roll.

6. **Open-ended** Name two events that are independent, and use them to illustrate the rule $P(A \text{ and } B) = P(A) \cdot P(B)$.

7. **Open-ended** Name two events that are mutually exclusive, and use them to illustrate the rule $P(A \text{ or } B) = P(A) + P(B)$.

8. Of the high school students surveyed, 65% of them reported that they preferred attending football games to attending soccer matches. Of the students surveyed, 45% were males. If gender is independent of sports preference, what is the probability that a randomly-chosen student is a male who prefers watching football to watching soccer?

Complementary events

If A and B are complementary events, then $P(A) = 1 - P(B)$. This rule is often useful for finding the probability of some event.

Example 2

A local television meteorologist says that in any given year there is a 20% chance of snowfall on New Year's Day in your area. What is the probability of at least one New Year's Day snowfall in a 10-year period?

■ Solution ■

There are two ways to find this probability. You could find the probabilities of exactly one New Year's Day snowfall, exactly two New Year's Day snowfalls, ..., and exactly ten New Year's Day snowfalls, and then add these probabilities together. (This would be very complicated.) Or you can use the fact that the complement of "at least one New Year's Day snowfall in a 10-year period" is "no New Year's Day snowfalls in a 10-year period."

(Solution continued on next page.)

According to the meteorologist, for any given New Year's Day there is a 100% – 20%, or 80% chance of no snowfall. (Remember, snowing and not snowing are complementary events.)

Since snowfall on each of the ten New Year's Days is an independent event, P(no New Year's Day snowfalls in a 10-year period) = $0.8 \cdot 0.8 \cdot 0.8 \cdot \cdots \cdot 0.8 = (0.8)^{10}$, or about 0.11.

Therefore, P(at least one New Year's Day snowfall in a 10-year period) = $1 - P$(no New Year's Day snowfalls in a 10-year period) $\approx 1 - 0.11$, or 0.89.

The probability of at least one New Year's Day snowfall in a 10-year period is about 0.89.

9. Refer to Example 2. What is the probability of snowfall on New Year's Day two years in a row? What is the probability of no snowfall on New Year's Day four years in a row?

10. What is the probability of obtaining at least one 6 in six rolls of a die?

11. What is the probability of a coin landing heads at least once in four flips of a coin?

12. Evandor plays baseball during the summer. His team has five games scheduled during the month of June. He estimates that the probability of rain on any particular day in June is 0.15. If there is rain on a game day, that day's baseball game will have to be rescheduled. What is the probability that at least one of his five June baseball games will have to be rescheduled due to rain?

13. An assembly line turns out about 3 defective items for every 100 items produced. What is the probability that at least one of the next 100 items is defective?

....
Spiral Review

14. One day at the library, Marita discovered that she has read 17 of the 65 books on one shelf of the science fiction section. What is the probability that Marita has read a randomly-selected book from the science fiction section? *(Section 13.1)*

Find the sum of each series. *(Sections 10.4 and 10.5)*

15. $9 + (-3) + 1 + \left(-\dfrac{1}{3}\right) + \cdots$

16. $\displaystyle\sum_{i=1}^{20} (2i - 3)$

13.3 Using Conditional Probability

Learn how to . . .
- find conditional probabilities

So you can . . .
- solve probability problems

Application

When drawing cards from a deck or selecting colored balls from a jar, it is necessary to know whether the items are drawn with or without replacement. If the first card or ball is returned before the second one is chosen, the outcomes of the two draws are independent. If the first item is not returned before the second one is chosen, the second outcome is dependent on the first and the situation involves conditional probability.

Terms to Know	**Example / Illustration**
Conditional probability (p. 626) the probability of event B given that event A has occurred	Suppose you draw two cards from a deck of 52 cards without replacement. Let A be the event that the first card drawn is a heart and let B be the event that the second card drawn is also a heart. $P(B \mid A) = \dfrac{12}{51}$, or $\dfrac{4}{17}$

UNDERSTANDING THE MAIN IDEAS

The conditional probability of an event B given that event A has occurred can be found using the formula $P(B \mid A) = \dfrac{P(A \text{ and } B)}{P(A)}$.

The previous section introduced the formula for finding the probability that two *independent* events will both occur: $P(A \text{ and } B) = P(A) \cdot P(B)$. The formula for finding conditional probability also provides us with a way to find the probability of both A and B occurring when A and B are not independent events. Multiplying both sides of the formula for conditional probability by $P(A)$ gives $P(A \text{ and } B) = P(B \mid A) \cdot P(A)$. (*Note*: This formula can be used for both independent and dependent events, since if A and B are independent events, then $P(B \mid A) = P(B)$.)

Example 1

An experiment consists of rolling a die once. Let A be the event the die shows a 1. Let B be the event the die shows an odd number. Find each of the following.

a. $P(B)$ **b.** $P(A \text{ and } B)$ **c.** $P(A \mid B)$

▪ Solution ▪

a. There are six possible outcomes and event B consists of three of them, so
$$P(B) = \frac{3}{6}, \text{ or } \frac{1}{2}.$$

b. Since the only outcome that is both 1 and odd is 1,
$$P(A \text{ and } B) = P(1) = \frac{1}{6}.$$

c. Using the results of parts (a) and (b), $P(A \mid B) = \dfrac{P(A \text{ and } B)}{P(B)} = \dfrac{\frac{1}{6}}{\frac{1}{2}} = \dfrac{1}{3}.$

Notice that part (c) of Example 1 is asking you to find the probability that the die shows 1 given that it shows an odd number. The result makes sense, since 1 is one of the three possible odd numbers.

A card is drawn at random from a standard deck of 52 cards. Let A be the event that the card is an ace, let B be the event that the card is a spade, and let C be the event that the card is black. Find each indicated probability.

1. $P(B)$ **2.** $P(C)$ **3.** $P(A \text{ and } B)$ **4.** $P(A \text{ and } C)$

5. $P(B \text{ and } C)$ **6.** $P(A \mid B)$ **7.** $P(B \mid C)$ **8.** $P(C \mid B)$

Some conditional probability problems can also be solved using a *probability tree diagram*.

Example 2

Jolon has 10 identical black socks and 14 identical brown socks in his dresser drawer. He reaches into the drawer and pulls out two socks, one sock at a time, without looking in the drawer. Find each probability.

a. the probability that he pulls out a matching pair if the first sock was black

b. the probability that he pulls out a matching pair if the first sock was brown

c. the probability that he pulls out a matching pair

■ Solution ■

a. You are asked to find P(Jolon got two matching socks | the first sock was black), which is the same as P(second sock is black | first sock is black). After drawing one black sock, there were 9 black and 14 brown socks remaining in the drawer.

So, P(second sock is black | first sock is black) $= \dfrac{9}{23} \approx 0.391$.

b. After drawing one brown sock, there were 10 black and 13 brown socks remaining in the drawer.

So, P(second sock is brown | first sock is brown) $= \dfrac{13}{23} \approx 0.565$.

c. This question is more complicated. It is helpful to construct a tree diagram representing Jolon's first and second picks. The probabilities of two of the second picks were found in parts (*a*) and (*b*) above.

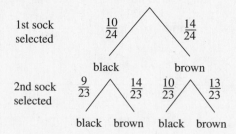

If Jolon pulls out a matching pair, it is either two brown socks or two black socks. These events are mutually exclusive. So,

$$P(\text{matching pair}) = P(\text{2 brown or 2 black})$$
$$= P(\text{2 brown}) + P(\text{2 black})$$
$$= \frac{10}{24} \cdot \frac{9}{23} + \frac{14}{24} \cdot \frac{13}{23} \qquad \leftarrow \text{Use the tree diagram.}$$
$$= \frac{45}{276} + \frac{91}{276}$$
$$\approx 0.493$$

The probability that Jolon draws a matching pair of socks is about 0.493.

In a certain city, there is a 60% chance that a day when it rains will be followed by rain the next day. There is a 30% chance that a day when it did not rain will be followed by rain the next day. A local meteorologist says that there is a 40% chance of rain in the city on Saturday.

9. Construct a probability tree diagram representing the chances for rain on Saturday and Sunday.

10. Use your tree diagram to find the probability it will rain both days this weekend.

11. What is the probability that it will rain at least one day this weekend?

A small factory has two assembly lines producing different products. Of the items produced, 70% of them are produced on assembly line A. On average, 3% of the output of assembly line A is defective and 5% of the output from assembly line B is defective.

12. Construct a tree diagram representing the output of the factory. (*Hint:* Show the assembly line an item comes from at the first branch, and whether it is good or defective at the second branch.)

13. What is the probability that a randomly-selected item produced by the factory is defective?

14. An item is selected at random from the factory's output and is found to be defective. What is the probability that it came from assembly line B?

A grocery store worker is giving out free samples of a new cracker, and offering customers a book of coupons for filling out a survey about it. The table below shows the shoppers' responses to a question about the likelihood of their purchasing the product. Use this information for Exercises 15–18.

Gender of shopper	Probably would buy	Probably would not buy
Male	25	10
Female	30	45

15. What proportion of the shoppers polled were male?

16. What is P(shopper would probably buy the product | shopper is male)?

17. According to the survey results, what proportion of shoppers would you expect to purchase the product?

18. What is P(shopper is male | shopper would probably buy the product)?

19. Mathematics Journal Explain how drawing cards from a deck without replacement is different from replacing each card and reshuffling the deck before drawing the next one. How do these procedures relate to conditional probability and independence?

········· · · · · · · · · ·
Spiral Review

Evaluate each logarithm. *(Section 4.2)*

20. $\log_7 49$

21. $\log \dfrac{1}{100}$

22. $\log_2 \dfrac{1}{64}$

23. $\log_6 \sqrt{6}$

Use the binomial theorem to expand each power of a binomial. *(Section 12.5)*

24. $(x - y)^5$

25. $(2x + y)^3$

26. $(s + 5)^4$

27. $(a - 3b)^4$

Binomial Distributions

Learn how to . . .

- find the probability distribution for a binomial experiment

So you can . . .

- make predictions in cases where there are many trials, each with two possible outcomes

Application

A quality control engineer knows that the output of a particular assembly line tends to be about 4% defective. He examines a random sample of 10 items and finds 2 of them to be defective. The probability of at least this many defectives, given that the overall proportion of defectives is still just 4%, is about 0.058. Probabilities of this type are found by using the binomial distribution.

Terms to Know

Example / Illustration

Binomial distribution (p. 634)
  a distribution of outcomes from an experiment in which there are only two mutually exclusive outcomes that are possible (The probability of k successes in n trials of the binomial experiment is given by the formula $_nC_k \cdot p^k(1-p)^{n-k}$.)

The number of defective items in a lot of 10 items from the assembly line discussed in the Application has a binomial distribution with $n = 10$ and $p = 0.04$.

UNDERSTANDING THE MAIN IDEAS

A *binomial experiment* is characterized by a sequence of independent trials, each of which can result in one of two mutually exclusive outcomes, usually called "success" and "failure." For each trial, the probability of a success is represented by p. (*Note:* If the probability of a succes is p, then the probability of a failure is $1 - p$.) For example, if the experiment is to roll a die 5 times and a success is obtaining a 1 or 6, then $p = P(1 \text{ or } 6) = \frac{1}{3}$. If the experiment consists of flipping a coin 3 times and a success is landing tails, then $p = \frac{1}{2}$.

Study Guide, ALGEBRA 2: EXPLORATIONS AND APPLICATIONS

Example

Lien is practicing free-throws at the basketball hoop in her driveway. She makes about 65% of her free-throws.

a. Find the binomial distribution for the number of free-throws she will make in her next 10 attempts.

b. What is the probability that she will make no more than 3 of her next 10 free-throws?

▪ Solution ▪

a. The number of free-throws she will make in her next 10 attempts has a binomial distribution with $n = 10$ and $p = 0.65$. Substitute these values for n and p in the formula $_nC_k \cdot p^k(1 - p)^{n - k}$. Using this formula for each of the values $k = 0, 1, 2, ..., 10$, and a calculator, gives these approximate results.

Number of successes (k)	0	1	2	3	4	5
Probability	3×10^{-5}	5×10^{-4}	0.0043	0.0212	0.0689	0.1536

Number of successes (k)	6	7	8	9	10
Probability	0.2377	0.2522	0.1757	0.0725	0.0135

b. Making "no more than 3" means making 0, 1, 2, or 3 of her next 10 free-throws. Using the table above,

$$P(\leq 3 \text{ successes}) = P(0) + P(1) + P(2) + P(3)$$
$$\approx (3 \times 10^{-5}) + (5 \times 10^{-4}) + 0.0043 + 0.0212$$
$$\approx 0.02603$$

The probability that Lien will make no more than 3 of her next 10 free throws is about 0.026, or 2.6%.

For Exercises 1–3, write a formula to describe the probability distribution for a binomial experiment with _n_ trials, each with probability _p_ of success. Then find the probability that the experiment will have exactly _k_ successful trials.

1. $n = 4, p = 0.4; k = 3$ **2.** $n = 7, p = 0.1; k = 6$ **3.** $n = 10, p = 0.3; k = 5$

4. Writing Explain why rolling a die eight times and counting the number of 3's is a binomial experiment. Write a formula that can be used to find the probability of k successes for $k = 0, 1, 2, ... , 8$.

---

5. Make a table of the probability distribution for the number of successful free-throws Rebecca makes in her next 10 attempts.

6. What is the probability that she will make at least 4 free-throws in her next 10 attempts?

Sonya has already collected cards for about 20% of the current major league baseball players. She usually buys cards in packets of 5.

7. Give the probability distribution for the number of duplicates (cards she already has) she will find in her next purchase of a five-card packet.

8. What is the probability that there are more than 3 duplicates in one five-card packet? 0 duplicates?

When Alvin is typing, he misspells about one of every 25 words by transposing some of the letters.

9. What is the probability that he can type a 30-word paragraph without a single mistake of this type?

10. What is the probability he will make fewer than 3 of these errors in a 30-word paragraph?

····················
Spiral Review

11. Suppose $P(A) = 0.5$ and $P(B \mid A) = 0.6$. Find $P(A \text{ and } B)$. *(Section 13.3)*

Find the mean and standard deviation of each set of data. *(Section 6.5)*

12. 56, 57, 119, 116, 126, 72, 184, 182, 80, 87

13. 3.00, 1.75, 1.90, 0.19, 1.80, 2.50, 2.30, 0.32

Normal Distributions

Application

Scores on the verbal and mathematics portions of the Scholastic Aptitude Test (SAT) are normally distributed with a mean of 500 and a standard deviation of 100. A score of 600, one standard deviation above the mean, is the 84th percentile. This means, for example, that a student who gets a score of 600 on the verbal section of the test has scored higher on this part of the test than about 84% of the other students taking the SAT.

Learn how to . . .

* recognize a normal distribution
* find probabilities involving normally distributed data

So you can . . .

* solve probability problems involving normal distributions

Terms to Know   *Example / Illustration*

Terms to Know	*Example / Illustration*
Normal distribution (p. 639) a distribution characterized by the bell-shaped curve (which is symmetric about its mean) that models the distribution	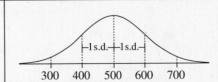 The curve above models a normal distribution with mean 500 and standard deviation 100.
Standard normal distribution (p. 641) the normal distribution with mean 0 and standard deviation 1	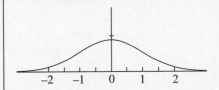
z-score (p. 641) $z = \dfrac{x - \bar{x}}{\sigma}$ where x is a value from a normal distribution that has mean $\bar{x}$ and standard deviation σ (The z-score gives the number of standard deviations that x lies from the mean.)	The z-score for a student who got a score of 550 on the mathematics portion of the SAT is $z = \dfrac{550 - 500}{100} = 0.5.$

UNDERSTANDING THE MAIN IDEAS

Characteristics of a normal distribution

A normal distribution is represented by a symmetric, bell-shaped curve, centered at its mean $\bar{x}$ with standard deviation σ. About 68% of the data lie within one standard deviation of $\bar{x}$, 95% within two standard deviations of $\bar{x}$, and 99% within three standard deviations of $\bar{x}$. The *area under the curve* (the area between the curve and the horizontal axis) of a normal distribution for a given interval on the horizontal axis gives an approximation of the percent of data that lies within that interval.

Example 1

The mean daytime high temperature during January in Rio de Janeiro, Brazil is about 79.4°F, with a standard deviation of 1.6°F. Assume that the temperatures are normally distributed.

a. During a typical January, what percent of days would you expect the high temperature to be between 77.8°F and 81°F?

b. Find the theoretical probability that the high temperature in Rio de Janeiro for a randomly-selected day in January would be less than 81°F.

■ Solution ■

a. First, you need to determine how many standard deviations the temperatures 77.8°F and 81°F are above or below the mean.

$$77.8°F - 79.4°F = -1.6°F \qquad 81°F - 79.4°F = 1.6°F$$

So, 77.8°F is one standard deviation *below* the mean and 81°F is one standard deviation *above* the mean.

Since the temperatures are normally distributed, about 68% of the high temperatures will lie within one standard deviation of the mean high temperature. So, for about 68% of the days in a typical January the daytime high temperature in Rio de Janeiro will be between 77.8°F and 81°F.

b. From part (a), 81°F is one standard deviation above the mean. Since a normal distribution is symmetric about the mean and 68% of the data lie within the interval from one standard deviation below the mean to one standard deviation above the mean, 0.5(68%) or 34% of the data lie in the interval from the mean to one standard deviation above the mean. By definition, half of the data lie below the mean. Thus, 34% of the temperatures lie between the mean temperature and 81°F, and 50% of the temperatures lie below the mean. So,

$$P(x < 81°F) = P(x < 79.4°F) + P(79.4°F < x < 81°F)$$
$$= 0.5 + 0.34$$
$$= 0.84$$

The theoretical probability that the high temperature in Rio de Janeiro for a randomly selected day in January would be less than 81°F is 0.84 (or 84%).

---

1. On about how many days in a typical January is the high temperature above 81°F in Rio de Janeiro?

2. Find an interval that would contain about 95% of the high temperature readings in Rio de Janeiro for the month of January.

The average daily high temperature during July in Rio de Janeiro is 69.9°F, with a standard deviation of 1.6°F.

3. During what percent of the days in a typical July is the high temperature in Rio de Janeiro between 66.7°F and 71.5°F?

4. Find an interval of temperatures that would contain about 99% of the daytime high temperature readings for July in Rio de Janeiro.

The scores on a history exam are normally distributed with mean 74 and standard deviation 8.

5. What percent of students had scores below 66?

6. What percent of students had scores above 90?

Standard normal distribution

The standard normal distribution is the normal distribution with mean 0 and standard deviation 1. A *standard normal table* can be used to find the probability that a randomly selected value from this distribution is less than a given number of standard deviations from the mean.

To find the probability that a value from a set of data that is normally distributed with mean $\bar{x}$ and standard deviation σ is less than some number x, you must first find the z-score for x using the formula $z = \dfrac{x - \bar{x}}{\sigma}$, in order to use the standard normal table. A standard normal table is shown in the Student Edition on page 641.

Example 2

The systolic blood pressures of 18-year-old females is normally distributed with a mean of 120 mm Hg and a standard deviation of 12 mm Hg. Find the probability that a randomly selected 18-year-old female will have each of these systolic readings.

   a. less than 140   **b.** between 112 and 140   **c.** over 150

■ Solution ■

To find each of these probabilities using the standard normal table, you convert each value into a z-score.

a. $P(x < 140) = P\left(z < \dfrac{140 - 120}{12}\right) \approx P(z < 1.7)$

From the standard normal table, $P(z < 1.7) \approx 0.9554$.

The probability that a randomly selected 18-year-old female will have a systolic blood pressure reading less than 140 mm Hg is about 0.9554.

b. $P(112 < z < 140) = P\left(\dfrac{112 - 120}{12} < z < \dfrac{140 - 120}{12}\right)$

$\approx P(-0.7 < z < 1.7)$

Since the standard normal table gives percentages *below* each z-score, the entry for $z = 1.7$ represents $P(z < 1.7)$ and the entry for $z = -0.7$ represents $P(z < -0.7)$. Therefore,

$P(-0.7 < z < 1.7) = P(z < 1.7) - P(-0.7 < z)$

$\approx 0.9554 - 0.2420$

≈ 0.7134

The probability that a randomly selected 18-year-old female will have a systolic blood pressure reading between 112 mm Hg and 142 mm Hg is about 0.7134.

c. $P(x > 150) = P\left(z > \dfrac{150 - 120}{12}\right)$

$= P(z > 2.5)$

Since the standard normal table gives percentages *below* each z-score, use the complementary probability.

$P(z > 2.5) = 1 - P(z < 2.5)$

$\approx 1 - 0.9938$

≈ 0.0062

The probability that a randomly selected 18-year-old female will have a systolic blood pressure reading over 150 mm Hg is about 0.0062.

Use the standard normal table to find each probability.

7. $P(z < -1.2)$ **8.** $P(z < 1.2)$ **9.** $P(z > 2.3)$ **10.** $P(0 < z < 1.4)$

Find the z-score corresponding to each value x if x comes from a normal distribution with mean $\bar{x}$ and standard deviation σ.

11. $\bar{x} = 5$; $\sigma = 4$; $x = 2$ **12.** $\bar{x} = 0$; $\sigma = 0.1$, $x = -0.2$

13. $\bar{x} = -3$; $\sigma = 0.5$; $x = -1.8$ **14.** $\bar{x} = 6.5$; $\sigma = 1.5$; $x = 2$

Study Guide, ALGEBRA 2: EXPLORATIONS AND APPLICATIONS

15. Refer to Example 2. What is the probability that a randomly selected 18 year old female will have a systolic blood pressure reading greater than 140 mm Hg? less than 100 mm Hg?

The number of miles Kyra Kendall drives in a given week is normally distributed with a mean of 145 mi and a standard deviation of 20 mi. Use this information for Exercises 16 and 17.

16. What percentage of weeks does Kyra drive more than 200 mi? fewer than 115 mi?

17. If Kyra budgets enough money for gas to drive 185 mi each week, what is the probability she will exceed her budget in any given week?

18. The amount of liquid a juice machine dispenses is normally distributed with a mean of 6.0 oz and a standard deviation of 0.4 oz. If more than 7.0 oz is poured into a cup it will overflow. About what percentage of the cups overflow when filled?

Scores on the mathematics portion of the SAT are normally distributed with a mean of 500 and a standard deviation of 100. Use this information for Exercises 19 and 20.

19. What score is at the 75th percentile, the percentile for which 75% of all scores are less than it? (*Hint:* Use the z-score corresponding to the entry closest to 0.75 in the standard normal table.)

20. What score is at the 90th percentile?

21. Mathematics Journal List the key characteristics of a normal distribution. Give some examples of data that tend to be normally distributed.

· · · · · · · · · · · · · · · · · · ·
Spiral Review

22. What is the probability that you will obtain exactly 3 heads in 5 flips of a fair coin? *(Section 13.4)*

23. Bob rides his bike to school, and on his route there is one intersection with a traffic light. He knows from experience that the light is red about 20% of the time. What is the probability that he will get stopped by the light exactly twice in one 5-day school week of commuting to and from school? *(Section 13.4)*

For Exercises 24–27, simplify each expression. *(Section 12.4)*

24. $_6C_0$ **25.** $_{11}C_2$ **26.** $_{10}C_3$ **27.** $_8C_4$

28. Are $\triangle ABC$ and $\triangle ACD$ similar? Explain. *(Toolbox, p. 801)*

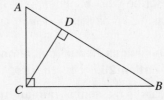

Complete these exercises for a review of Chapter 13. If you have difficulty with a particular problem, review the indicated section.

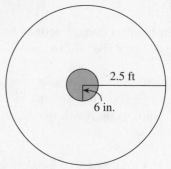

1. Karl attends summer camp each year, and tries out a new sport on each visit. This year he decides to take up the sport of archery. When he begins, he just concentrates on holding the bow correctly, lining up the arrow with the target, and getting off a clean shot. He is equally likely to hit any point on the target. At this stage, what is the probability that he will get a bull's-eye if his arrow hits the target? What kind of probability is this? *(Section 13.1)*

2. After several days of lessons, Karl's aim has improved considerably. Today he hit the bull's-eye 4 out of his first 11 attempts. What is the probability that he will hit the bull's-eye with his next arrow? What kind of probability is this? *(Section 13.1)*

An experiment consists of selecting a card from an ordinary deck of 52 playing cards, noting what it is, returning it to the deck and reshuffling, then selecting another card and noting its identity as well. Let *A* be the event that the first card is a spade, let *B* be the event that the first card is red, and let *C* be the event that the second card is black. Find each probability. *(Section 13.2)*

3. $P(A \text{ and } B)$ 4. $P(A \text{ or } B)$ 5. $P(A \text{ and } C)$ 6. $P(B \text{ and } C)$

An urn contains 5 red balls, 6 yellow balls, and 4 green balls. Two balls are selected at random, one after the other, without replacement. *(Section 13.3)*

7. What is the probability that the first ball is red?

8. Find the probability that the second ball is green given that the first is red.

9. What is the probability that the first ball is red and the second ball is green?

When Anna plays table tennis with her brother Charlie, the probability that she wins a point on one of her serves is about 0.7. For Exercises 10 and 11, find each probability. *(Section 13.4)*

10. She wins exactly 1 point in her next five serves.

11. She wins 4 points in her next five serves.

12. **Writing** If $p = 0.5$, the histogram for the probabilities of a binomial distribution is symmetric. Describe what happens to the histogram as the value of p approaches 0 and as it approaches 1. *(Section 13.4)*

The annual rainfall in El Seybo, Dominican Republic is normally distributed with mean 138.8 cm and standard deviation 25.4 cm. *(Section 13.5)*

13. During what percentage of years is the annual rainfall in El Seybo between 88 cm and 215 cm?

14. During what percentage of years is the annual rainfall in El Seybo less than 113.4 cm?

The weights of two-year-old girls are normally distributed with mean 11.9 kg and standard deviation 1.28 kg. Use this information for Exercises 15 and 16. *(Section 13.5)*

15. Find an interval that contains the middle 68% of two-year-old girls' weights.

16. What is the probability that a randomly selected two-year-old girl weighs more than 12.5 kg?

17. Open-ended Suppose the scores on an exam are normally distributed with mean 75 and standard deviation 8. Decide what percentage of students should receive A's, B's, and so on, and determine the interval of scores for each letter grade. *(Section 13.5)*

SPIRAL REVIEW   Chapters 1–13

For Exercises 1 and 2, rewrite each equation in standard form. Then identify the conic associated with the equation.

1. $9y^2 - 16x^2 + 128x + 36y = 76$ **2.** $4x^2 + 8x - y + 2 = 0$

3. Write a matrix equation equivalent to the system of equations shown at the right. Then find the solution.

$$5x - 2y = -4$$
$$3x - 4y = 6$$

For Exercises 4 and 5, find the sum of each series, if it exists.

4. $\displaystyle\sum_{i=1}^{8} (2i - 3)$ **5.** $4 - 2 + 1 - \dfrac{1}{2} + \cdots$

6. Write out the 8th row of Pascal's triangle. How are these numbers related to the probabilities for success in a binomial experiment where $n = 8$ and $p = 0.4$?

7. Solve $\log_2 x + \log_2 (x - 3) = 2$.

8. Write an equation for y as a function of x if y varies inversely as the square of x and $y = 6$ when $x = 2$.

9. Find all the zeros of the polynomial function $f(x) = x^4 - 2x^3 - 2x^2 - 2x - 3$.

10. The force exerted by a spring is directly proportional to the distance it is displaced from its equilibrium position. If a spring exerts a force of 5 lb when it is displaced 2.25 in., what force will it exert when it is displaced 3 in. in the same direction?

11. Give the domain and range of the function $y = \sqrt{x + 2} - 1$, and sketch its graph.

12. To get home from school, Katie walks 0.4 mi due south, turns due east and continues walking for 0.7 mi. What is the straight-line distance between her school and her house?

14.1

Using Sine and Cosine

Learn how to . . .

- find the sine and cosine of an acute angle

So you can . . .

- find the length of a leg of a right triangle

Application

For safety reasons, an extension ladder must form an angle with the ground ($\angle A$) no greater than 80°. You can use the sine and cosine ratios to relate the length of the ladder (AB) the distance between the foot of the ladder and the wall (AC), and the distance from the base of the wall to the top of the ladder (BC).

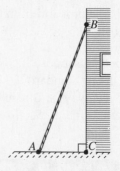

Terms to Know

Example / Illustration

Sine (p. 656)

the ratio of the length of the leg opposite an angle and the length of the hypotenuse in a right triangle

$$\sin A = \frac{\text{length of side opposite } \angle A}{\text{length of hypotenuse}}$$

Cosine (p. 656)

the ratio of the length of the leg adjacent to an angle and the length of the hypotenuse in a right triangle

$$\cos A = \frac{\text{length of side adjacent to } \angle A}{\text{length of hypotenuse}}$$

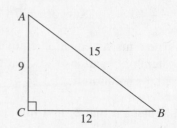

$$\sin A = \frac{12}{15} = \frac{4}{5} \qquad \sin B = \frac{9}{15} = \frac{3}{5}$$

$$\cos A = \frac{9}{15} = \frac{3}{5} \qquad \cos B = \frac{12}{15} = \frac{4}{5}$$

UNDERSTANDING THE MAIN IDEAS

Finding two missing sides in a right triangle

If you know the measure of one of the acute angles in a right triangle and the length of the hypotenuse or a leg of the triangle, then you can use the sine or cosine ratio to find the lengths of the other two sides of the triangle.

Example 1

For each triangle, write a ratio for the sine and cosine of each acute angle. Then find the lengths of the missing sides of the triangle.

a.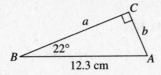

b.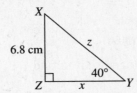

■ Solution ■

a. $\sin A = \dfrac{\text{opposite}}{\text{hypotenuse}} = \dfrac{a}{12.3}$ $\cos A = \dfrac{\text{adjacent}}{\text{hypotenuse}} = \dfrac{b}{12.3}$

$\sin B = \dfrac{\text{opposite}}{\text{hypotenuse}} = \dfrac{b}{12.3}$ $\cos B = \dfrac{\text{adjacent}}{\text{hypotenuse}} = \dfrac{a}{12.3}$

To find a use $\sin A$ or $\cos B$.

$$\cos 22° = \frac{a}{12.3}$$

$$a = 12.3 \cos 22°$$

$$\approx (12.3)(0.9272) \quad \leftarrow \text{Use a calculator.}$$

$$\approx 11.4 \text{ cm}$$

To find b use $\sin B$ or $\cos A$.

$$\sin 22° = \frac{b}{12.3}$$

$$b = 12.3 \sin 22°$$

$$\approx (12.3)(0.3746) \quad \leftarrow \text{Use a calculator.}$$

$$\approx 4.61 \text{ cm}$$

(Solution continues on next page.)

■ **Solution** ■ *(continued)*

b. $\sin X = \dfrac{\text{opposite}}{\text{hypotenuse}} = \dfrac{x}{z}$ $\qquad \cos X = \dfrac{\text{adjacent}}{\text{hypotenuse}} = \dfrac{6.8}{z}$

$\sin Y = \dfrac{\text{opposite}}{\text{hypotenuse}} = \dfrac{6.8}{z}$ $\qquad \cos Y = \dfrac{\text{adjacent}}{\text{hypotenuse}} = \dfrac{x}{z}$

First find z, then use it to find x.

$\sin 40° = \dfrac{6.8}{z}$ $\qquad \leftarrow \angle Y = 40°$

$z = \dfrac{6.8}{\sin 40°}$

$\approx \dfrac{6.8}{0.6428}$

≈ 10.6 cm

$\cos 40° = \dfrac{x}{z}$

$\approx \dfrac{x}{10.6}$ $\qquad \leftarrow z \approx 10.6$

$x = 10.6 \cos 40°$

$\approx (10.6)(0.7660)$

≈ 8.12 cm

Write a ratio for sin *A* and cos *A* for each triangle.

1.

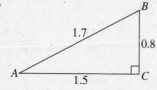

2.

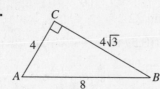

Use a calculator to find each value.

3. $\sin 30°$ **4.** $\cos 30°$ **5.** $\sin 45°$

6. $\cos 45°$ **7.** $\sin 0°$ **8.** $\cos 0°$

Find the missing lengths in each right triangle.

9.

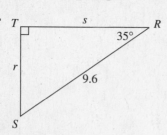

10.

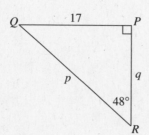

Expressing fractions of a degree

You can express the measure of an angle in degrees, minutes, and seconds, or as a decimal value.

Example 2

a. Express 53°14′57″ (read "53 degrees, 14 minutes, 57 seconds") in decimal degrees, to the nearest tenth of a degree.

b. Express 3.892° in degrees, minutes, and seconds, to the nearest second.

■ Solution ■

a. One minute is $\frac{1}{60}$ of a degree and one second is $\frac{1}{3600}$ of a degree.

So $53°14′57″ = \left(53 + \frac{14}{60} + \frac{57}{3600}\right)^{\circ}$

$\approx 53 + 0.2333 + 0.015833)^{\circ}$

$\approx 53.2°$

(*Note*: Many calculators will convert between angle measures in degrees, minutes, and seconds (DMS) and decimal degrees (DD).)

b. Use the decimal part of 3.892° to find the number of minutes. Then use the decimal part of the number of minutes to find the number of seconds.

$3.892° = 3° + (0.892)(60′)$

$= 3° + 53.52′$

$= 3° + 53′ + (0.52)(60″)$

$= 3° + 53′ + 31.2″$

$\approx 3°53′32″$

Express each angle measure in decimal degrees, to the nearest tenth of a degree.

11. 27°53′35″

12. 79°12′9″

Express each angle measure in degrees, minutes, and seconds, to the nearest second.

13. 83.012°

14. 10.897°

Find the measure of an acute angle in a right triangle

If you know the length of a leg and the hypotenuse of a right triangle, you can find the measures of the acute angles of the triangle. Finding such angle measures involves using the inverse sine or inverse cosine function.

Example 3

Find the measure of $\angle A$ in $\triangle ABC$. Express the angle measure in decimal degrees, to the nearest tenth of a degree, and in degrees, minutes, and seconds, to the nearest second.

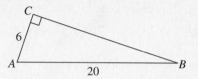

■ Solution ■

First find the cosine ratio for $\angle A$. Then use the inverse cosine function on a calculator to find the angle measure in decimal degrees.

$$\cos A = \frac{\text{adjacent}}{\text{hypotenuse}} = \frac{6}{20} = 0.3$$

To find $\angle A$, use the inverse cosine function on a calculator.

$$\angle A = \cos^{-1} 0.3 \approx 72.5° \qquad \leftarrow \text{ Use the COS}^{-1} \text{ key or the INV COS keys.}$$

To write the measure of $\angle A$ in degrees, minutes, and seconds, use the technique of Example 2 or a calculator that does the conversion.

$$\angle A \approx 72°32'33''$$

The measure of $\angle A$ is about $72.5°$ or about $72°32'33''$.

For each trigonometric ratio, find θ. Express the answer in two ways:
a. in decimal degrees, to the nearest tenth of a degree.
b. in degrees, minutes, and seconds, to the nearest second.

15. $\sin \theta = 0.4210$

16. $\cos \theta = 0.9501$

17. $\sin \theta = \dfrac{3}{5}$

18. $\cos \theta = \dfrac{7}{11}$

Find the measure of $\angle A$ in decimal degrees, to the nearest tenth of a degree.

19.

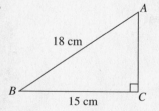

20.

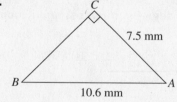

A 40-foot ladder rests against a building and forms an angle of 80° with the ground.

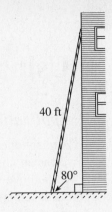

21. How far is the foot of the ladder from the side of the building?

22. How high up the side of the building does the ladder reach?

....
Spiral Review

23. Suppose a track team's practice hours form a normal distribution. If the mean length of a practice is 2.75 h and the standard deviation is 0.5 h, what percent of the practice sessions have been longer than 3.25 h? shorter than 1.75 h? *(Section 13.5)*

Find the geometric mean for each pair of numbers. *(Section 10.2)*

24. 14, 9

25. 2.7, 0.13

26. $\dfrac{7}{12}, \dfrac{5}{8}$

Solve each proportion. *(Toolbox, p. 785)*

27. $\dfrac{y}{3} = \dfrac{14}{8}$

28. $\dfrac{9}{z} = \dfrac{z}{10}$

29. $\dfrac{5}{7} = \dfrac{20}{5x + 3}$

GOAL

Using Tangent

Learn how to . . .

- find the tangent of an acute angle

So you can . . .

- find all the missing sides and angles of a right triangle

Application

A telephone pole on level ground is to be supported by three wires. Each wire forms an angle of 52° with the ground, and the wires are attached to the pole 40 ft above the ground. You can use the tangent ratio to calculate the proper distance between the base of the pole and the point where each wire is anchored to the ground.

Terms to Know	*Example / Illustration*
Tangent (p. 663) the ratio of the length of the leg opposite an angle and the length of the leg adjacent to the angle in a right triangle $\tan A = \dfrac{\text{length of side opposite } \angle A}{\text{length of side adjacent to } \angle A}$	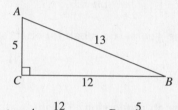 $\tan A = \dfrac{12}{5} \qquad \tan B = \dfrac{5}{12}$
Angle of depression (p. 665) when looking down at an object, the angle whose sides are your line of sight and a horizontal line through your viewpoint	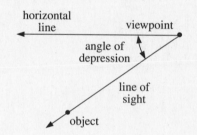
Angle of elevation (p. 665) when looking up at an object, the angle whose sides are your line of sight and a horizontal line through your viewpoint	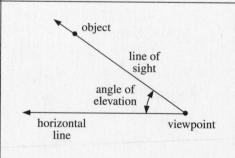

352

Terms to Know	Example / Illustration
Solving a triangle (p. 655) finding the measures of *all* sides and angles of a triangle	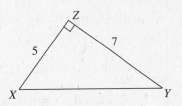 Given the lengths of any two sides of a right triangle, you can find length of the third side and the measures of the two acute angles. (See Example 1a.) 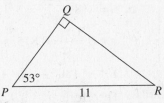 Given the length of any side and the measure of either acute angle of a right triangle, you can find the lengths of the other two sides and the measure of the other acute angle. (See Example 1c.)

UNDERSTANDING THE MAIN IDEAS

Solving triangles

When solving a right triangle, use any of the three trigonometric ratios (sine, cosine, and tangent) and their inverse functions, the Pythagorean theorem, and the fact that the sum of the angles of any triangle is 180°.

Example 1

Solve $\triangle ABC$ using the given measures.

  a. $a = 5$, $b = 7$ **b.** $a = 7$, $c = 8$

  c. $\angle B = 53°$, $c = 11$ **d.** $\angle A = 23°$, $a = 5$

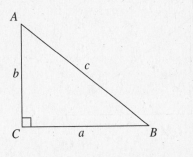

■ Solution ■

a. Using the Pythagorean theorem:

$$c = \sqrt{5^2 + 7^2} = \sqrt{74} \approx 8.6$$

Using the tangent ratio and the inverse tangent function:

$$\tan B = \frac{7}{5} = 1.4, \text{ so } \angle B = \tan^{-1}(1.4) \approx 54.5° \quad \leftarrow \text{Use a calculator.}$$

Since the sum of the measures of a triangle is 180°,

$$\angle A \approx 180° - (90° + 54.5°) = 35.5°$$

Therefore, $c \approx 8.6$, $\angle A \approx 35.5°$, and $\angle B \approx 54.5°$.

b. $b = \sqrt{8^2 - 7^2} = \sqrt{15} \approx 3.9$

$$\cos B = \frac{7}{8} = 0.875, \text{ so } \angle B = \cos^{-1}(0.875) \approx 29.0°$$

$$\angle A \approx 180° - (90° + 29.0°) = 61.0°$$

Therefore, $b \approx 3.9$, $\angle B \approx 29.0°$, and $\angle A \approx 61.0°$.

c. $\angle A = 180° - (90° + 53°) = 37°$

$$\sin 53° = \frac{b}{11}, \text{ so } b = 11 \sin 53° \approx (11)(0.7986) \approx 8.8$$

$$\cos 53° = \frac{a}{11}, \text{ so } a = 11 \cos 53° \approx (11)(0.6018) \approx 6.6$$

Therefore, $\angle A = 37°$, $b \approx 8.8$, and $a \approx 6.6$.

d. $\angle B = 180° - (90° + 23°) = 67°$

$$\tan 23° = \frac{5}{b}, \text{ so } b = \frac{5}{\tan 23°} \approx \frac{5}{0.4245} \approx 11.8$$

$$\sin 23° = \frac{5}{c}, \text{ so } c = \frac{5}{\sin 23°} \approx \frac{5}{0.3907} \approx 12.8$$

Therefore, $\angle B = 67°$, $b \approx 11.8$, and $c \approx 12.8$.

Write a ratio for tan A and tan B for each right triangle. Then find the measures of ∠A and ∠B to the nearest tenth of a degree.

1.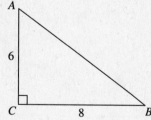

2.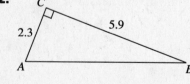

Solve each triangle.

3.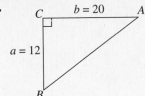
$b = 20$
$a = 12$

4.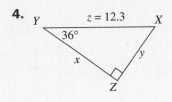
$z = 12.3$
$36°$
x
y

5.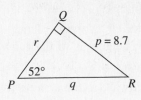
r
$52°$
$p = 8.7$
q

Angles of elevation and depression

When you look up, the angle formed by your line of sight and a horizontal line is called an *angle of elevation*. When you look down, the angle formed by your line of sight and a horizontal line is called an *angle of depression*.

Example 2

While flying at an altitude of 600 ft, a falcon sees a gopher on the ground at an angle of depression of 28°. At the same time, a field mouse sees the falcon at an angle of elevation of 22°.

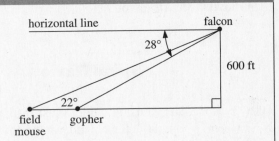
horizontal line   falcon
28°
600 ft
22°
field mouse   gopher

a. At that moment, what is the distance between the falcon and the gopher?

b. At that moment, what is the distance between the falcon and the field mouse?

■ Solution ■

a. Using $\triangle GAF$ at the right, you need to find a.
Use the sine function.

$$\sin 28° = \frac{600}{a}$$

$$a = \frac{600}{\sin 28°} \approx \frac{600}{0.4695} \approx 1278 \text{ ft}$$

The gopher is about 1278 ft from the falcon.

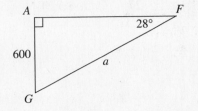
A F
28°
600
a
G

b. Using $\triangle MBF$ at the right, you need to find b.
Use the sine function.

$$\sin 22° = \frac{600}{b}$$

$$b = \frac{600}{\sin 22°} \approx \frac{600}{0.3746} \approx 1602 \text{ ft}$$

The field mouse is about 1602 ft from the falcon.

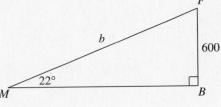
F
b
600
22°
M B

As a pilot begins to land her airplane, she notes that the angle of depression to the beginning of the runway is 32°25′.

6. If the altitude of the airplane is 5400 ft at this time, how far is the airplane from the beginning of the runway, as measured along the ground?

7. If the angle of depression to the other end of the runway is 24°15′, how long is the runway (to the nearest foot)?

For Exercises 8 and 9, refer to the Application at the beginning of this section.

8. What is the distance between the base of the pole and the point where each support wire is anchored?

9. What is the length of each wire?

10. **Mathematics Journal** Suppose you are on the ground and you see that an airplane, cruising at a level altitude, is flying toward you. Explain what happens to the angle of elevation of the airplane as it approaches, flies over, and then moves away from you.

....
Spiral Review

11. For safety reasons, an extension ladder must form an angle with the ground that is between 72° and 80°. If the ladder is 55 ft long and the top rests against the side of a building, what is the range for the distance between the wall and the foot of the ladder in order to meet this safety requirement? *(Section 14.1)*

12. State whether each sequence is *geometric* or *arithmetic*. State the common difference or common ratio for each sequence. *(Section 10.2)*

    a. 18, 15, 12, 9, ...   **b.** 18, 15, 12.5, ...

13. Name the vertex and the line of symmetry for the graph of the equation $y = -3x^2 - 12x - 27$. *(Section 5.2)*

Angles of Rotation

Application

Using sonar, a salvage ship can determine the angle θ and distance r to a sunken vessel. Then the crew can use the values of θ and r to calculate the depth of y of the sunken vessel and the distance x from the ship to a point directly above it.

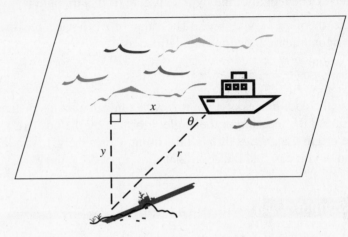

Terms to Know	*Example / Illustration*
Initial side of an angle (p. 671) the non-rotated side of an angle formed by rotating a ray in a counterclockwise direction from the positive x-axis (This side of the angle remains on the positive x-axis.)	
Terminal side of an angle (p. 671) the rotated side of an angle formed by rotating a ray in a counterclockwise direction from the positive x-axis	
Standard position of an angle (p. 671) an angle with its terminal side on the positive x-axis and its vertex at the origin	The angle shown above is in standard position.

UNDERSTANDING THE MAIN IDEAS

Using $\sin\theta = \dfrac{y}{r}$, $\cos\theta = \dfrac{x}{r}$, and $\tan\theta = \dfrac{y}{x}$

The trigonometric functions of an angle θ between 0° and 360° are defined by placing θ in standard position in a coordinate plane and choosing a point $P(x, y)$ (other than the origin) on the terminal side of the angle. By drawing a line segment from P that is perpendicular to the x-axis, a right triangle is formed in which the "lengths" of the legs are x and y. (These "lengths" can be positive, negative, or zero.) The length of the hypotenuse, r, of this triangle is given by the Pythagorean theorem as $\sqrt{x^2 + y^2}$. (The value of r is always positive.) The values of the trigonometric functions are defined as

$$\sin\theta = \frac{y}{r}, \quad \cos\theta = \frac{x}{r}, \quad \text{and} \quad \tan\theta = \frac{y}{x}.$$

So if you know the coordinates of a point P, you can calculate its distance r from the origin and the measure of the angle θ formed by the positive x-axis and a ray from the origin that passes through the point. Conversely, if you know the values of r and θ, you can find the coordinates (x, y) of the point P.

Example 1

For each point P, find the distance r between the point and the origin. Then find $\sin\theta$, $\cos\theta$, and $\tan\theta$ for an angle θ in standard position, with the given point P on its terminal side.

a. $P(-4, 7)$ **b.** $P(-3, -5)$ **c.** $P(2, -6)$

■ Solution ■

a. $r = \sqrt{(-4)^2 + 7^2} = \sqrt{65}$ ← $x = -4$ and $y = 7$

$\sin\theta = \dfrac{y}{r} = \dfrac{7}{\sqrt{65}} \approx 0.8682$

$\cos\theta = \dfrac{x}{r} = \dfrac{-4}{\sqrt{65}} \approx -0.4961$

$\tan\theta = \dfrac{y}{x} = \dfrac{7}{-4} = -\dfrac{7}{4}$

b. $r = \sqrt{(-3)^2 + (-5)^2} = \sqrt{34}$

$\sin\theta = \dfrac{y}{r} = \dfrac{-5}{\sqrt{34}} \approx -0.8575$

$\cos\theta = \dfrac{x}{r} = \dfrac{-3}{\sqrt{34}} \approx -0.5145$

$\tan\theta = \dfrac{y}{x} = \dfrac{-5}{-3} = \dfrac{5}{3}$

(Solution continues on next page.)

■ Solution *(continued)* ■

c. $r = \sqrt{2^2 + (-6)^2} = \sqrt{40} = 2\sqrt{10}$

$\sin \theta = \dfrac{y}{r} = \dfrac{-6}{2\sqrt{10}} \approx -0.9487$

$\cos \theta = \dfrac{x}{r} = \dfrac{2}{2\sqrt{10}} \approx 0.3162$

$\tan \theta = \dfrac{y}{x} = \dfrac{-6}{2} = -3$

Example 2

Find the coordinates of the point P associated with each pair of values r and θ.

a. $r = 10, \ \theta = 240°$ **b.** $r = 1, \ \theta = 175°$

■ Solution ■

a. Use the cosine ratio to find the value of x.

$\cos \theta = \dfrac{x}{r}$

$\cos 240° = \dfrac{x}{10}$

$x = 10 \cos 240°$

$= 10(-0.5)$

$= -5$

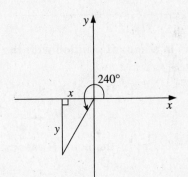

Use the sine ratio to find the value of y.

$\sin \theta = \dfrac{y}{r}$

$\sin 240° = \dfrac{y}{10}$

$y = 10 \sin 240°$

$= 10\left(-\dfrac{\sqrt{3}}{2}\right)$

$= -5\sqrt{3}$

The coordinates of point P are $(-5, -5\sqrt{3})$.

(Solution continues on next page.)

b. Use the cosine ratio to find the value of x.

$$\cos \theta = \frac{x}{r}$$

$$\cos 175° = \frac{x}{1}$$

$$x = \cos 175°$$

$$\approx -0.9962$$

Use the sine ratio to find the value of y.

$$\sin \theta = \frac{y}{r}$$

$$\sin 175° = \frac{y}{1}$$

$$x = \sin 175°$$

$$\approx 0.0872$$

The coordinates of point P are about $(-0.9962, 0.0872)$.

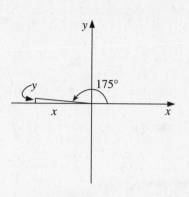

An angle θ is in standard position with the given point P on its terminal side. Find sin θ, cos θ, and tan θ.

1. $P(-3, 4)$ **2.** $P(-5, -12)$ **3.** $P(6, -6\sqrt{3})$

4. $P(-7, -7)$ **5.** $P(1, 10)$ **6.** $P(-15, 30)$

Find the coordinates of the point P associated with each pair of values r and θ.

7. $r = 5$, $\theta = 60°$ **8.** $r = 3$, $\theta = 150°$ **9.** $r = 2$, $\theta = 180°$

10. $r = 7.8$, $\theta = 297°$ **11.** $r = 0.5$, $\theta = 205°$ **12.** $r = 1$, $\theta = 135°$

Finding angles of rotation

For any given value of sin θ, cos θ, or tan θ, there are two possible angle measures for θ.

In the figure at the right, notice that $\sin A = \frac{y}{r} = \frac{5}{\sqrt{26}} = \sin B$.

In general, $\sin \theta = \sin (180° - \theta)$. Also, $\cos \theta = \cos (360° - \theta)$ and $\tan \theta = \tan (180° + \theta)$.

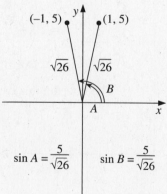

Also, the value of the trigonometric functions can be positive or negative depending on the quadrant in which the terminal side of the angle lies. The value of sin θ is positive in Quadrants I and II, and negative in Quadrants III and IV. The value of cos θ is positive in Quadrants I and IV, and negative in Quadrants II and III. The value of tan θ is positive in Quandrants I and III, and negative in Quadrants II and IV.

Example 3

Find the angles θ between 0° and 360° having each given trigonometric value.

a. $\sin \theta = -0.5$ **b.** $\cos \theta = 0.35$ **c.** $\tan \theta = -8.2$

■ Solution ■

a. The value of $\sin \theta$ is negative in Quadrants III and IV. Using the inverse sine function on a calculator,

$$\theta = \sin^{-1}(-0.5) = -30°$$

Since negative angle measures indicate clockwise rather than counterclockwise rotations, −30° is equivalent to 330°. Also, since $\sin \theta = \sin(180° - \theta)$, the other angle measure is $180° - (-30°)$, or 210°.

There are two angle measures between 0° and 360° whose sine ratio is −0.5, $\theta = 210°$ and $\theta = 330°$.

b. The value of $\cos \theta$ is positive in Quadrants I and IV. Using the inverse cosine function on a calculator,

$$\theta = \cos^{-1}(0.35) \approx 69.5°$$

Since $\cos \theta = \cos(360° - \theta)$, the other angle measure is approximately $360° - (69.5°)$, or 290.5°.

There are two angle measures between 0° and 360° whose cosine ratio is 0.35, $\theta \approx 69.5°$ and $\theta \approx 290.5°$.

c. The value of $\tan \theta$ is negative in Quadrants II and IV. Using the inverse tangent function on a calculator,

$$\theta = \tan^{-1}(-8.2) \approx -83.0°$$

Since negative angle measures indicate clockwise rather than counterclockwise rotations, −83.0° is equivalent to 277.0°. Also, since $\tan \theta = \tan(180° + \theta)$, the other angle measure is approximately $180° + (-83.0°)$, or 97.0°.

There are two angle measures between 0° and 360° whose tangent ratio is −8.2, $\theta = 277.0°$ and $\theta = 97.0°$.

Find all angles θ between 0° and 360° having each given trigonometric value. Give answers to the nearest tenth of a degree.

13. $\sin \theta = 0.6$ **14.** $\cos \theta = -0.25$ **15.** $\tan \theta = 0.95$

16. $\sin \theta = -0.65$ **17.** $\cos \theta = 0.325$ **18.** $\tan \theta = -2.05$

Refer to the Application at the beginning of this section. A ship's sonar indicates that a sunken vessel is 1540 m from the ship, at an angle of depression of 62°.

19. At what depth is the sunken vessel?

20. How far from the ship is the point directly above the sunken vessel?

....
Spiral Review

21. Solve each triangle. *(Section 14.2)*

a.

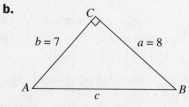

b.

22. Find the area of each triangle described. *(Toolbox, p. 802)*

a. base = 12 m; height = 8 m   **b.** base = 8 cm; height = $8\sqrt{10}$ cm

23. Suppose you roll a six-sided die four times. What is the probability of getting at least one even number? *(Section 13.2)*

Finding the Area of a Triangle

Learn how to . . .

- find the area of a triangle given two side lengths and the measure of the included angle

- find the area of a sector

So you can . . .

- solve problems involving the areas of triangles or sectors

Application

A company is surveying a triangular plot of level land. The survey team has found the measure of the angle formed at one corner of the property and the lengths of two sides of the property that meet at that corner. They can now use a formula to calculate the area of the triangular plot.

UNDERSTANDING THE MAIN IDEAS

A formula for the area of a triangle

If you know the lengths of two sides of any triangle and the measure of the angle they form, you can calculate the area of the triangle.

For $\triangle ABC$ shown at the right, the area K is found using one of the following formulas.

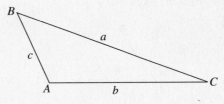

$$K = \frac{1}{2} bc \sin A \qquad K = \frac{1}{2} ac \sin B \qquad K = \frac{1}{2} ab \sin C$$

(*Note*: These formulas are valid for both acute and obtuse angle measures.)

Example 1

Find the area of each triangle.

a.

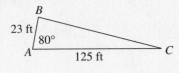

b.

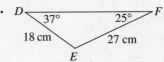

■ Solution ■

a. Use the formula $K = \frac{1}{2} bc \sin A$ with $b = 125$, $c = 23$, and $A = 80°$.

$$K = \frac{1}{2}(23)(125) \sin 80°$$

$$\approx \frac{1}{2}(23)(125)(0.9848)$$

$$\approx 1415.65$$

The area of the triangle is about 1416 ft².

b. The two sides whose lengths are given form $\angle E$, whose measure is not given. Since the sum of the measures of the angles of a triangle is 180°, $\angle E = 180° - (37° + 25°)$, or 118°.

Use the formula $K = \frac{1}{2} df \sin E$ with $d = 27$, $f = 18$, and $E = 118°$.

$$K = \frac{1}{2}(27)(18) \sin 118°$$

$$\approx \frac{1}{2}(27)(18)(0.8829)$$

$$\approx 214.5447$$

The area of the triangle is about 215 ft².

For Exercises 1–8, find the area of △DEF. If there is not enough information, explain.

1.

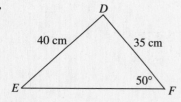

2.

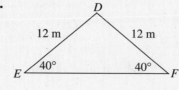

3.

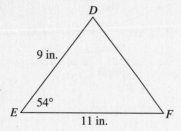

4.

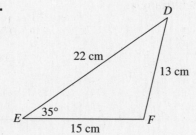

5. $e = 7$ in., $f = 8$ in., $\angle D = 87°$ **6.** $e = 2.5$ m, $f = 3.9$ m, $\angle F = 32°$

7. $f = 38$ m, $e = 46$ m, $\angle D = 110°$ **8.** $d = 22$ cm, $f = 22$ cm, $\angle E = 60°$

A formula for the area of a sector

In general, for any sector of a circle having radius r and central angle θ:

$$\frac{\text{area of the sector}}{\text{area of the circle}} = \frac{\theta}{360}$$

$$\text{area of the sector} = \frac{\theta}{360} \cdot (\text{area of the circle})$$

So, to find the area of a sector of a circle, find the product $\frac{\theta}{360} \cdot \pi r^2$ where r is the radius of the circle.

Example 2

Find the area of each sector having central angle θ and radius r.

   a. $\theta = 60°$, $r = 10$ cm   **b.** $\theta = 310°$, $r = 5$ in.

■ Solution ■

a. Use the formula *area of sector* $= \frac{\theta}{360} \cdot \pi r^2$ where $\theta = 60$ and $r = 10$.

$$\text{area of the sector} = \frac{60}{360} \cdot \pi(10)^2$$

$$= \frac{1}{6}(100\pi)$$

$$\approx 52.36$$

The area of the sector is about 52.4 cm^2.

b. area of sector $= \frac{\theta}{360} \cdot \pi r^2$ $\leftarrow$ Substitute 310 for θ and 5 for r.

$$= \frac{310}{360} \cdot \pi(5)^2$$

$$= \frac{31}{36}(25\pi)$$

$$\approx 67.63$$

The area of the sector is about 67.6 in.^{2}.

For Exercises 9–12, find the area of each shaded sector.

9.

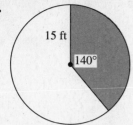

10.

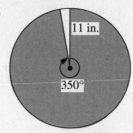

11.

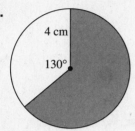

12.

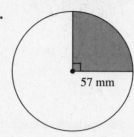

13. Writing Explain why the two formulas $A = \frac{1}{2} bh$ and $K = \frac{1}{2}xy \sin Z$ give the same result for the area of the triangle shown at the right.

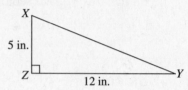

14. Refer to the Application at the beginning of this section. Suppose the survey team found that the lengths of two sides of the triangular plot are 12.36 m and 47.95 m, and that the angle formed by these two sides measured 110°22′47″, what is the area of the plot (to the nearest hundredth of a square meter)?

15. Open-ended If you cut a sector from a circular disk as shown at the right, you can join the cut edges and form a cone. By experimenting with several disks, determine how the size of ∠AOB affects the height of the cone and the angle at the tip of the cone.

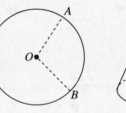

....
Spiral Review

Find two measures of ∠A for each trigonometric value. *(Section 14.3)*

16. $\sin A = 0.4553$

17. $\tan A = -1.235$

18. $\cos A = -0.6251$

Tell whether y varies inversely with x. If so, state the constant of variation. *(Section 9.6)*

19. $xy = -3$

20. $y = \dfrac{0.7}{x}$

21. $y = \dfrac{x}{13}$

22. $x = \dfrac{6}{y}$

14.5

The Law of Sines

Learn how to . . .

- use the law of sines to solve a triangle

So you can . . .

- solve problems involving the sides and angles of a triangle

Application

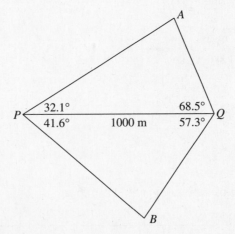

One surveying method is to start with a known baseline, such as $\overline{PQ}$ in the figure above, and measure the angles from the endpoints of this baseline (P and Q) to the points of interest (A and B). Then the law of sines can be used to calculate the distances between the endpoints and the points of interest, PA, QA, PB, and QB in the figure.

UNDERSTANDING THE MAIN IDEAS

You can use the law of sines to solve a triangle when you know the measures of two angles of a triangle and the length of one of the sides. The law of sines states:

for any $\triangle ABC$, $\dfrac{\sin A}{a} = \dfrac{\sin B}{b} = \dfrac{\sin C}{c}$.

Example 1

Find the length of $\overline{AQ}$ in $\triangle APQ$.

■ Solution ■

In order to use the law of sines, first find the measure of $\angle A$.

$$\angle A = 180° - (32.1° + 68.5°) = 79.4°$$

Now use the law of sines.

$$\frac{\sin A}{a} = \frac{\sin P}{p}$$

$$\frac{\sin 79.4°}{1000} = \frac{\sin 32.1°}{p}$$

$$p = \frac{1000 \sin 32.1°}{\sin 79.4°}$$

$$\approx \frac{1000(0.5314)}{0.9829}$$

$$\approx 540.6$$

The length of side $\overline{AQ}$ is about 540.6 m.

---

1. For $\triangle ABC$, $\angle A = 87°$, $\angle B = 62°$, and $a = 35$. Find the length of $\overline{AC}$.

2. For $\triangle DEF$, $\angle F = 76°$, $d = 13$, and $f = 29$. Find the measure of $\angle D$.

3. For $\triangle GHJ$, $\angle H = 28°$, $\angle J = 102°$, and $h = 32$. Find the length of $\overline{GH}$.

4. For $\triangle KLM$, $\angle L = 38°$, $\angle M = 42°$, and $m = 1.32$. Find the length of $\overline{LM}$.

Determining the number of triangles

The situation in which two sides and a non-included angle of a triangle are known is called the *ambiguous case*, because it does not always determine a unique triangle. When you know the lengths of two sides and the measure of a non-included angle of a triangle, you can use the law of sines once you know how many triangles are possible. If the known angle measure is that of an acute angle, there can be 0, 1, or 2 triangles possible, depending on the known side lengths. If the known angle measure is that of an obtuse angle, there can be 0 or 1 triangles possible.

Example 2

The measure of one angle and the lengths of two sides of $\triangle ABC$ are given. Tell the number of triangles that are determined and solve each possible triangle. If no triangle is possible, explain why not.

  a. $\angle C = 35°$, $b = 17$ m, $c = 10$ m

  b. $\angle B = 60°$, $c = 15$ cm, $b = 5$ cm

  c. $\angle A = 115°$, $b = 4$ in., $a = 10$ in.

■ Solution ■

a. A sketch shows that two triangles are possible. First solve the triangle in which ∠B an is acute angle.

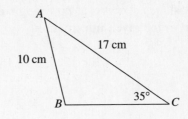

Using the law of sines:

$$\frac{\sin 35°}{10} = \frac{\sin B}{17}$$

$$\sin B = \frac{17 \sin 35°}{10}$$

$$\approx \frac{(17)(0.5736)}{10}$$

$$= 0.97512$$

$$\angle B \approx \sin^{-1}(0.9751)$$

$$\approx 77.2°$$

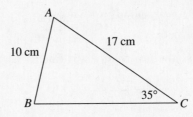

So ∠A ≈ 180° − (35° + 77.2°) = 67.8°.

Now use the law of sines again to find the length of $\overline{BC}$.

$$\frac{\sin 67.8°}{a} = \frac{\sin 35°}{10}$$

$$a = \frac{10 \sin 67.8°}{\sin 35°}$$

$$\approx \frac{(10)(0.9259)}{0.5736}$$

$$\approx 16.142$$

In acute △ABC, ∠A ≈ 67.8°, ∠B ≈ 77.2°, and a ≈ 16.1 cm.

The diagram at the right shows that the measure of obtuse ∠B is the supplement of the measure of acute ∠B.

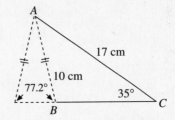

So ∠B ≈ 180° − 77.2° = 102.8°.

Then ∠A ≈ 180° − (102.8° + 35°) = 42.2°.

$$\frac{\sin 42.2°}{a} = \frac{\sin 35°}{10}$$

$$a = \frac{10 \sin 42.2°}{\sin 35°}$$

$$\approx \frac{(10)(0.6717)}{0.5736}$$

$$\approx 11.710$$

In obtuse △ABC, ∠A ≈ 42.2°, ∠B ≈ 102.8°, and a ≈ 11.7 cm.

(Solution continues on next page.)

■ **Solution** ■ *(continued)*

b. A sketch shows that no triangle can be formed from the given information.

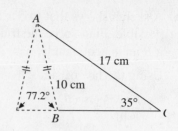

c. A sketch shows that one triangle can be formed.

Use the law of sines to find $\angle B$.

$$\frac{\sin B}{4} = \frac{\sin 115°}{10}$$

$$\sin B = \frac{4 \sin 115°}{10}$$

$$\approx \frac{(4)(0.9063)}{10}$$

$$= 0.36252$$

$$\angle B \approx \sin^{-1}(0.3625)$$

$$\approx 21.3°$$

So $\angle C \approx 180° - (115° + 21.3°) = 43.7°$.

Use the law of sines to find the length of $\overline{AB}$.

$$\frac{\sin 43.7}{c} = \frac{\sin 115°}{10}$$

$$c = \frac{10 \sin 43.7°}{\sin 115°}$$

$$\approx \frac{(10)(0.6909)}{0.9063}$$

$$\approx 7.623$$

In $\triangle ABC$, $\angle B \approx 21.3°$, $\angle C \approx 43.7°$, and $c \approx 7.6$ in.

The following tests can also be used to determine the number of triangles when two sides and a non-included angle of a triangle are known.

If the known angle ($\angle A$) is acute:

- No triangle is possible if $a < b \sin A$.
- One triangle is possible if $a = b \sin A$.
- Two triangles are possible if $b \sin A < a < b$.
- One triangle is possible if $a > b$.

If the known angle ($\angle A$) is obtuse:

- No triangle is possible if $a < b$.
- One triangle is possible if $a > b$.

Solve each triangle. If there are two solutions, give both. If there is no solution, explain why.

5.

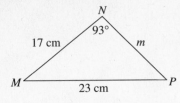

6.

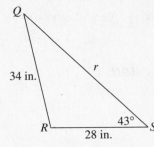

7.

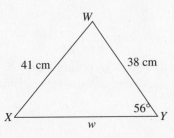

8.

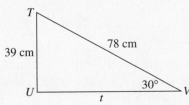

9. $\angle A = 135°, b = 14, a = 23$

10. $\angle A = 72°, c = 23, a = 18$

11. $\angle A = 65°, c = 43, a = 41$

12. $\angle A = 55°, c = 7, a = 12$

For Exercises 13–15, use the triangle shown at the right.

13. Use the law of sines to find the measure of $\angle J$.

14. Find the measure of $\angle G$.

15. Use the formula $K = \frac{1}{2}ab \sin C$ to find the area of the triangle.

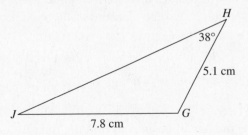

16. Mathematics Journal Refer to the discussion following Example 2. One of the tests identifies a triangle that must be a right triangle. Which test is it? Write a few sentences explaining why the triangle must be a right triangle.

....
Spiral Review

Find the area of △PQR. *(Section 14.4)*

17. $p = 7, q = 9, \angle R = 80°$

18. $\angle P = 35°, q = 15, r = 10$

19. $p = n, q = n, \angle R = 60°$

20. $\angle Q = 45°, r = 7, p = 7\sqrt{2}$

Write as a logarithm of a single number or expression. *(Section 4.3)*

21. $\frac{1}{2} \log_k 16$

22. $\log_n 8 - \log_n 16$

23. $r \log 9^q$

24. $5 \ln p^3 + 3 \ln p^6$

The Law of Cosines

Learn how to . . .

- use the law of
  cosines to solve a
  triangle

So you can . . .

- solve problems
  involving sides and
  angles of a triangle

Application

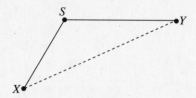

A ship at point S can use sonar to determine its distances to points X and Y. By measuring $\angle S$ and using the law of cosines, the crew can calculate the distance between X and Y.

UNDERSTANDING THE MAIN IDEAS

You can use the law of cosines to solve a triangle when you know the lengths of two sides of a triangle and the measure of the included angle. The law of cosines states:

for any $\triangle ABC$, $\qquad a^2 = b^2 + c^2 - 2bc \cos A$

$$b^2 = a^2 + c^2 - 2ac \cos B$$

$$c^2 = a^2 + b^2 - 2ab \cos C$$

You can also use the law of cosines to find the measures of the angles of a triangle when you know the lengths of all three sides of the triangle. Solving each of the three equations above for $\cos A$, $\cos B$, and $\cos C$, respectively:

for any $\triangle ABC$, $\qquad \cos A = \dfrac{b^2 + c^2 - a^2}{2bc}$

$$\cos B = \dfrac{a^2 + c^2 - b^2}{2ac}$$

$$\cos C = \dfrac{a^2 + b^2 - c^2}{2ab}$$

Once the value of the expression on the right side of the three equations above has been computed, the measure of the angle can be found by taking the inverse cosine of both sides.

a. Find the missing side length e for $\triangle DEF$ below.

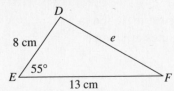

b. Find the measure of $\angle R$ in $\triangle PQR$ below.

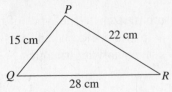

■ Solution ■

a. The lengths of two sides of the triangle and the measure of the angle between them are given, so you can use the law of cosines to find the length of the side opposite the known angle.

$$e^2 = d^2 + f^2 - 2df \cos E$$
$$= 13^2 + 8^2 - 2(13)(8) \cos 55° \quad \leftarrow d = 13, f = 8, E = 55°$$
$$\approx 169 + 64 - 208(0.5736)$$
$$\approx 133.6912$$
$$x \approx \sqrt{133.6912}$$
$$\approx 11.56$$

The length e is about 11.6 cm.

b. The lengths of the three sides of the triangle are given, so you can use the law of cosines to find the measure of $\angle R$.

$$\cos R = \frac{p^2 + q^2 - r^2}{2pq}$$
$$= \frac{28^2 + 22^2 - 15^2}{2(28)(22)}$$
$$= \frac{1043}{1232}$$
$$\approx 0.8466$$
$$\angle R \approx \cos^{-1}(0.8466)$$
$$\approx 32.2°$$

The measure of $\angle R$ is about 32.2°.

Find the missing side length of each triangle.

1.

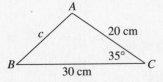

2.

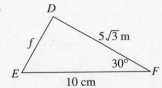

3.

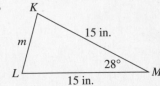

Find the angle measures of each triangle.

4. $\triangle XYZ$ with $x = 26$, $y = 24$, $z = 17$
5. $\triangle XYZ$ with $x = 15$, $y = 12$, $z = 9$
6. $\triangle XYZ$ with $z = 14$, $y = 23$, $x = 18$
7. $\triangle XYZ$ with $x = 3\sqrt{2}$, $y = 3$, $z = 3$

The following chart summarizes all the types of problems discussed in Sections 14.5 and 14.6. It can help you decide when to use the law of cosines and when to use the law of sines when solving triangles.

Given information:	Law to use:
two angles and a non-included side	law of sines (First find the measure of the included angle.)
two sides and a non-included angle	law of sines (First decide if there are 0, 1, or 2 triangles.)
two angles and the included side	law of sines (First find the measure of the third angle.)
two sides and the included angle	law of cosines
three sides	law of cosines

For Exercises 8–13, tell which of the situations from the chart above is represented by each triangle. Then solve the triangle.

8.
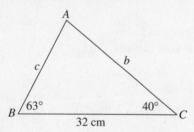

9.
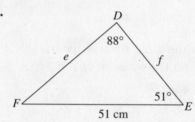

10.
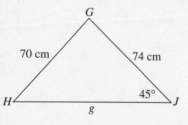

11.
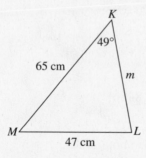

12.
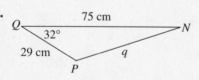

13.
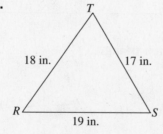

14. Refer to the Application at the beginning of this section. Suppose point X is an island 3800 ft from the ship, point Y is a floating bottle (with a note inside) that is 2750 ft from the ship, and $\angle S = 145°$. If the bottle is drifting toward the island at a constant rate of 8 feet per minute, how long will it take the bottle (in hours, minutes, and seconds) to wash up on the shore of the island?

15. Mathematics Journal Explain why you cannot solve a triangle if the only given measures are those of the three angles of the triangle.

· · · · · · · · · · · · · · · · · · · ·
Spiral Review

Find each missing side length and angle measure of $\triangle ABC$. *(Section 14.5)*

16. $c = 3$, $\angle A = 87°$, $\angle C = 35°$ **17.** $b = 13$, $\angle B = 79°$, $\angle C = 52°$

Find sin θ and cos θ for each value of θ. *(Section 14.3)*

18. $\theta = 175°$ **19.** $\theta = 42°$ **20.** $\theta = 250°$

21. $\theta = 350°$ **22.** $\theta = 192°$ **23.** $\theta = 270°$

Chapter 14 Review ·············

Complete these exercises for a review of Chapter 14. If you have difficulty with a particular problem, review the indicated section.

Write a ratio for the sine, cosine, and tangent of each acute angle. Then use a calculator to find the measure of each acute angle to the nearest degree. *(Sections 14.1 and 14.2)*

1.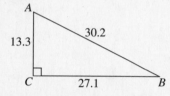

2.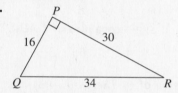

Express each angle measure in degrees, minutes, and seconds, to the nearest second. *(Section 14.1)*

3. $65.357°$

4. $130.027°$

For Exercises 5 and 6, express each angle measure in decimal degrees, to the nearest tenth of a degree. *(Section 14.1)*

5. $138°15'27''$

6. $2°3'4''$

7. An office worker in one of two identical office buildings notices that from her office the angle of elevation to the roof of the other building is $37°10'$, and that the angle of depression to its base is $50°23'$. If the buildings are 35 m apart, how tall are they? *(Section 14.2)*

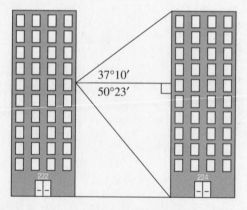

An angle θ is in standard position with the given point P on its terminal side. Find sin θ, cos θ, and tan θ. *(Section 14.3)*

8. $(8, -11)$

9. $(-3.7, -4.9)$

Find the coordinates of the point P associated with each pair of values r and θ. *(Section 14.3)*

10. $r = 17, \theta = 115°$

11. $r = 22.5, \theta = 295°$

Find all values of θ between $0°$ and $360°$ having each given trigonometric value. Give answers to the nearest tenth of a degree. *(Section 14.3)*

12. $\sin \theta = 0.8511$

13. $\cos \theta = -0.7201$

14. $\tan \theta = -1.9535$

376

Find the area of each shaded region. *(Section 14.4)*

15.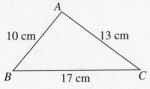

16.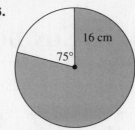

Solve each triangle. If there are two solutions, give both. If there is no solution, explain why. *(Sections 14.5 and 14.6)*

17.

18.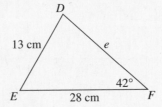

19.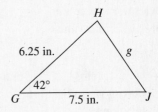

Find each product or quotient.

1. $(5 + 3i)(7 - 12i)$

2. $\dfrac{2 - 3i}{5 + 6i}$

3. $(5x^3 + 2x^2 - 3x + 5)(2x - 1)$

4. $(8x^4 - 26x^3 + 3x^2 + 17x - 6) \div (4x - 3)$

Solve each system of equations.

5. $3x - 5y = 13$
$-7x + 2y = 18$

6. $2x - 3y + 4z = 3$
$3x - 5y - 2z = 22$
$5x + 2y - 3z = 5$

For Exercises 7 and 8, use the quadratic formula to solve each equation.

7. $2x^2 - 15x - 3 = 0$

8. $(x + 1)(x - 7) = 2x + 3$

9. Line *n* is parallel to the line $3x + 2y = 8$, and goes through the midpoint of the segment with endpoints (1, 7) and (–5, 3). Write an equation for line *n*.

10. What is the equation of the line through the points (1, 7) and (–5, 3)?

11. What is the equation of the circle that has the points (1, 7) and (–5, 3) as the endpoints of a diameter?

The first four terms of a sequence are 15, 11, 7, 3.

12. What are the next three terms?   **13.** What is the sum of the first ten terms?

Five cards are selected from a deck. They are the 4 of diamonds, the 6 of hearts, the 6 of spades, the Jack of clubs, and the ace of hearts.

14. How many different ways can you stack the cards from top to bottom?

15. How many different ways can you stack the cards if you start and end with a 6?

Sine and Cosine on the Unit Circle

Learn how to . . .

- work with sine and cosine as functions
- extend the domain of sine and cosine

So you can . . .

- solve problems involving periodic phenomena

Application

In an AC electric generator, electricity is produced in coils of wire wound around a rotating core called the *armature*. The "push" of the electricity, or voltage, depends on the angle of the armature relative to a magnet.

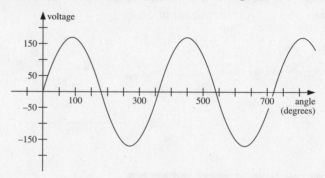

The voltage *v* is given as a function of the angle *t* by the equation

$$v = 120\sqrt{2} \sin t.$$

Terms to Know

Example / Illustration

Sine (p. 708) the vertical position of a point on the unit circle as a function of the angle its radius makes with the positive *x*-axis	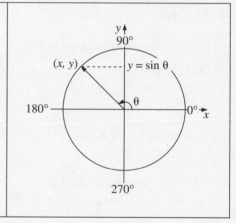

Terms to Know	Example / Illustration
Cosine (p. 708) the horizontal position of a point on the unit circle as a function of the angle its radius makes with the positive *x*-axis	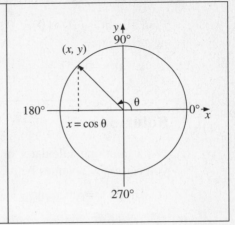

UNDERSTANDING THE MAIN IDEAS

Sine and cosine

The sine and cosine functions relate lengths to angles. As a point moves counterclockwise around the unit circle, the value of its sine (its vertical position with respect to the *x*-axis) first increases from 0 to 1, then decreases back to 0, then decreases further from 0 to –1, before finally increasing back to its initial value, 0.

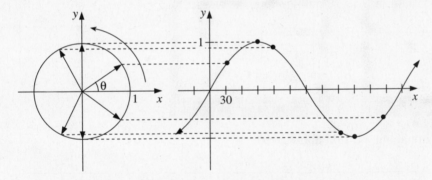

A similar picture could be used to illustrate the cosine function.

While movement around the unit circle in a counterclockwise direction is associated with positive angle measures, movement in a clockwise direction is associated with negative angle measures. Also, unless explicitly stated, these movements are not restricted to a single revolution around the circle but can be more than one revolution. Therefore, the domain of the sine and cosine functions is any angle measure, positive or negative.

Example

Find all angles θ for $0° \leq \theta \leq 360°$ that satisfy each equation. Give answers to the nearest degree.

a. $\sin \theta = -0.5237$

b. $\cos \theta = 0.7548$

■ Solution ■

Use a graphing calculator or graphing software set in degree mode. Set the window so you can view x-values between 0 and 360.

a. Graph $y = \sin x$ and $y = -0.5237$, and find the intersections of the two graphs.

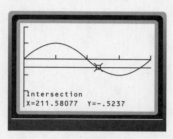
Intersection
X=211.58077 Y=-.5237

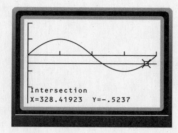
Intersection
X=328.41923 Y=-.5237

The two angles are about 212° and about 328°.

b. Graph $y = \cos x$ and $y = 0.7548$, and find the intersections of the two graphs.

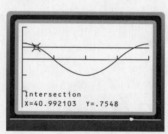
Intersection
X=40.992103 Y=.7548

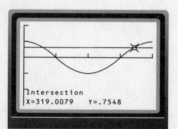
Intersection
X=319.0079   Y=.7548

The two angles are about 41° and about 319°.

Find all angles θ for $0° \leq \theta \leq 360°$ that satisfy each equation. Give answers to the nearest degree.

1. $\sin \theta = 0.6253$

2. $\cos \theta = 0.3287$

3. $\sin \theta = -0.2239$

4. $\cos \theta = -0.5407$

5. $\sin \theta = 0.6581$

6. $\cos \theta = 0$

Evaluate each expression.

7. $\sin 140°$

8. $\cos 39°$

9. $\sin 233°$

10. $\cos (-15°)$

The voltage v of an electric generator is given by the function $v = 120\sqrt{2} \sin \theta$, where θ is the angle of the armature. Find the voltage for each value of θ. Use this information for Exercises 11–14.

11. $45°$

12. $165°$

13. $24°$

14. $105°$

---

15. Open-ended Describe some real-life situations that might be modeled by a sine function or a cosine function.

Spiral Review

For Exercises 16–19, solve each △ABC. *(Section 14.6)*

16. $a = 11$, $b = 14$, $c = 17$ 　　　　　 **17.** $\angle A = 43°$, $\angle C = 75°$, $b = 12$

18. $b = 23$, $c = 18$, $\angle A = 110°$ 　　　　 **19.** $a = 5.6$, $b = 9.2$, $c = 8.1$

20. A circle has a circumference of 8 cm. *(Toolbox, page 802)*

 a. What is the radius of the circle?

 b. An arc of the circle is 2 cm long. What is the measure of the central angle that intercepts the arc?

Find the area of each shaded region. *(Section 14.4)*

21.
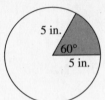
5 in.
60°
5 in.

22.
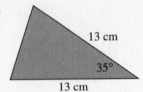
13 cm
35°
13 cm

23.
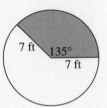
7 ft
135°
7 ft

Measuring in Radians

Learn how to . . .

- convert between degree measure and radian measure
- find the sine and cosine of angles given in the radians

So you can . . .

- explore models of periodic phenomemena

Application

As a bicycle wheel moves forward, the *angle* through which the wheel has rotated can be measured by marking off the distance the wheel has traveled in radius lengths.

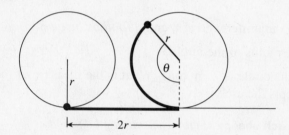

The wheel has rotated through an angle θ of 2 radians.

Terms to Know	**Example / Illustration**
Radian (p. 714) the measure of an angle expressed as the length of the unit circle's arc intercepted by the angle	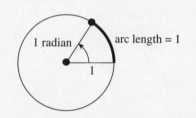 The measure of this angle is 1 radian. 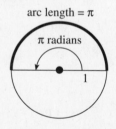 The measure of this angle is π radians (≈ 3.14 radians).

UNDERSTANDING THE MAIN IDEAS

Radian measure

In degrees, one complete revolution around the unit circle represents an angle measure of 360°. Since the circumference of the unit circle is 2π, one complete revolution around the unit circle also represents an angle measure of 2π radians. Therefore, 2π radians = 360°.

Dividing both sides of this equality by 2π shows that

$$1 \text{ radian} = \frac{180°}{\pi}$$

Similarly, dividing the both sides of the equality by 360 shows that

$$1° = \frac{\pi}{180} \text{ radians}$$

(*Note*: If an angle measure does not have a degree symbol, it is understood to be in radians.)

Example 1

a. Convert $\frac{5\pi}{6}$ radians to degrees.   **b.** Convert 149° to radians.

■ Solution ■

a. Since 1 radian $= \frac{180°}{\pi}$, multiply $\frac{5\pi}{6}$ by $\frac{180°}{\pi}$.

$$\frac{5\pi}{6} \text{ radians} = \frac{5\pi}{6} \cdot \frac{180°}{\pi} = \frac{5 \cdot 180°}{6} = 150°$$

b. Since $1° = \frac{\pi}{180}$ radians, multiply 149 by $\frac{\pi}{180}$.

$$149 = 149 \cdot \frac{\pi}{180} \text{ radians} \approx 2.6 \text{ radians}$$

Convert each angle measure from radians to degrees (to the nearest degree).

1. $\frac{\pi}{2}$ **2.** 2π **3.** $\frac{2\pi}{3}$ **4.** $-\frac{7\pi}{6}$

5. 2.3   **6.** 4.8   **7.** -3.7 **8.** 1.4

Convert each angle measure from degrees to radians. Express each answer as a decimal to the nearest hundredth of a radian, and in terms of π.

9. 50°   **10.** 240°   **11.** $-144°$ **12.** 54°

13. 315°   **14.** $-110°$ **15.** 80°   **16.** 222°

---

17. A wheel of a mountain bike has a radius of 1 foot. Find the distance the bike travels when the wheel rotates through angles of 40°, 160°, 420°, and 750°.

Sine and cosine of angles given in radians

To find the sine or cosine of an angle given in radians, locate the terminal side of the angle by means of the unit circle, or use a calculator set to radian mode.

Example 2

a. Find $\cos \dfrac{3\pi}{2}$.

b. Find $\sin 4.2$.

■ Solution ■

a.

The arc length $\dfrac{3\pi}{2}$ is $\dfrac{3}{4}$ of 2π, the circumference of the unit circle. So $\cos \dfrac{3\pi}{2}$ is the x-coordinate of the point $(0, -1)$.

$$\cos \dfrac{3\pi}{2} = 0$$

b. Use a calculator. Make sure the calculator is set in radian mode.

$\sin 4.2 \approx -0.8716$

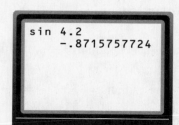

```
sin 4.2
        -.8715757724
```

For Exercises 18–25, evaluate each expression.

18. $\sin \dfrac{\pi}{2}$ **19.** $\cos \dfrac{5\pi}{4}$ **20.** $\sin \dfrac{\pi}{6}$ **21.** $\cos \left(-\dfrac{\pi}{6}\right)$

22. $\cos 1.8$ **23.** $\sin 2.5$ **24.** $\cos 4.7$ **25.** $\sin (-3.6)$

26. Mathematics Journal Describe how you could find the sine of an angle given a diagram in which the angle is drawn in a circle that is not the unit circle.

For Exercises 27–29, find all angles θ for $0° \leq \theta \leq 360°$ that satisfy each equation. Give answers to the nearest degree. *(Section 15.1)*

27. $\cos \theta = 0.7071$ **28.** $\sin \theta = 0.4731$ **29.** $\cos \theta = -0.8660$

30. Use a graphing calculator or graphing software to graph the function $y = -2 + 3 \sin \theta$. *(Section 15.1)*

    a. What is the maximum value of the function?

    b. What is the minimum value of the function?

31. Suppose you choose a card from among all the red cards of a standard deck. What is the probability of getting the 5 or 6 of diamonds? *(Section 13.1)*

Exploring Amplitude and Period

Learn how to . . .

- graph equations of these forms:
  $y = a \sin bx + k$
  $y = a \cos bx + k$

So you can . . .

- understand periodic phenomena

Application

In two cities that are located on the same longitude line, such as St. Cloud, Minnesota, and Port Arthur, Texas, the seasons change at the same time. However, since St. Cloud is farther north, the variation in the number of hours of daylight each day over a calendar year is greater. If you graph the number of hours of daylight against the months of the year for the two cities, the two graphs have the same *period*, but they have different *amplitudes*.

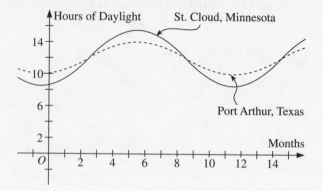

Terms to Know | **Example / Illustration**

Cycle (p. 720) the smallest portion of the graph of a periodic function that repeats	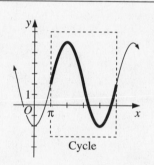
Period (p. 720) the length of one cycle of the graph of a periodic function	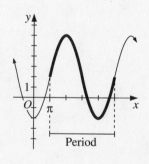

386

Amplitude (p. 720)

the number $\dfrac{M-m}{2}$, where M is the maximum value and m is the minimum value of the function

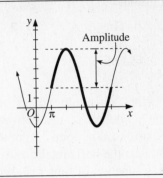

UNDERSTANDING THE MAIN IDEAS

Period and amplitude

The amplitude of a graph whose equation is of the form $y = a \sin bx$ or $y = a \cos bx$ is $|a|$. The period of such a graph is $\dfrac{2\pi}{|b|}$, and its axis (horizontal center line) is the x-axis.

Example 1

Graph each equation.

a. $y = 3 \cos \dfrac{\pi x}{2}$ **b.** $y = -2 \sin 3x$

Solution

a. The amplitude of the graph is $|3| = 3$, so the graph rises 3 units above and falls 3 units below the x-axis.

The period of the graph is $\dfrac{2\pi}{|\pi/2|} = 4$, so the graph completes one cycle every 4 units. The graph is shown at the right.

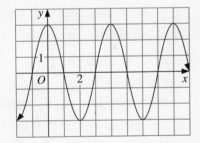

b. The amplitude is $|-2| = 2$, so the graph rises 2 units above and falls 2 units below the x-axis.

The period of the graph is $\dfrac{2\pi}{|3|} = \dfrac{2\pi}{3}$, so the graph completes one cycle every $\dfrac{2\pi}{3}$ units. Because the value of a is negative, the graph is falling as it passes through the origin. The graph is shown at the right.

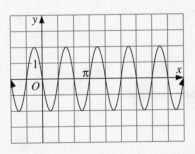

For Exercises 1–6, graph each function. State the amplitude and period of each graph.

1. $y = \sin 2x$

2. $y = 3 \cos 2\pi x$

3. $y = 2 \sin \dfrac{\pi}{4} x$

4. $y = -5 \cos \dfrac{2}{3} x$

5. $y = -4 \sin \dfrac{\pi}{3} x$

6. $y = 6 \cos \dfrac{3}{2} x$

7. A clock's pendulum clock makes one back-and-forth swing in 1 s. The tip of the pendulum swings between a point 15 cm to the right of its center position and a point 15 cm to the left of its center position. A function that models the horizontal position x (in centimeters) of the tip of the pendulum as a function of time t (in seconds) is $x = a \cos bt$. Find the values of a and b for the pendulum described.

Vertical translation

As shown in the following example, adding a constant k to a sine or cosine function moves the graph up or down, depending on whether k is positive or negative. Instead of the axis of the graph being the x-axis, the axis is the horizontal line $y = k$. Also, the value of k is the average of M and m, the maximum and minimum values of the function, respectively.

Example 2

Graph each equation on one set of axes.

a. $y = \sin \dfrac{3\pi}{2} x$

b. $y = \sin \dfrac{3\pi}{2} x + 2$

c. $y = \sin \dfrac{3\pi}{2} x - 1$

■ Solution ■

The graph of each function is a sine curve with amplitude 1 and period $\dfrac{4}{3}$. the graphs differ by the equation of their axes.

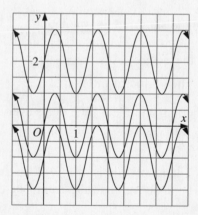

  a. The axis of the graph is the x-axis; it is the middle graph in the figure.

  b. The axis of the graph is the line $y = 2$; it is the top graph in the figure.

  c. The axis of the graph is the line $y = -1$; it is the bottom graph in the figure.

Graph each equation.

8. $y = 1 + 2 \sin 3x$

9. $y = -3 + \cos \dfrac{3}{2} x$

10. $y = 4 + 3 \sin \dfrac{x}{2}$

Write a cosine function with the given maximum M, minimum m, and period.

11. $M = 4$, $m = -6$, period $= \dfrac{1}{2}$

12. $M = 3.5$, $m = 2.5$, period $= \dfrac{\pi}{3}$

A ferris wheel with a radius of 30 ft has an axle that is 36 ft above the ground. When running, the wheel makes one complete turn in 80 s. Suppose you are in the highest seat when the wheel begins to turn counterclockwise.

13. Draw a graph of your height above the ground as a function of time.

14. Write an equation for your height above the ground as a function of time.

·····················
Spiral Review

Convert each angle measure from radians to degrees. *(Section 15.2)*

15. $\dfrac{\pi}{12}$

16. -3.5π

17. $\dfrac{7\pi}{4}$

Multiply. *(Section 8.4)*

18. $(3 - i)(5 + 7i)$

19. $(4 - 9i)(4 + 9i)$

20. $(-2 + 5i)^2$

Graph each function. *(Section 9.7)*

21. $y = \dfrac{4}{x + 3}$

22. $y = \dfrac{12}{x - 4} + 2$

23. $y = \dfrac{-3}{x + 1} - 4$

Exploring Phase Shifts

Learn how to ...

- graph equations of these forms:
$$y = a \sin b(x - h) + k$$
$$y = a \cos b(x - h) + k$$

- write equations for sine and cosine graphs that have a phase shift

So you can ...

- describe periodic phenomena that have a phase shift

Application

Because the planet Earth takes time to absorb the heat it receives from the sun, the changes in ground temperature lag behind the corresponding changes in air temperature. This *phase shift* is the reason that the warmest time of year is not at the *beginning* of summer, when average daily heat from the sun reaches its highest level.

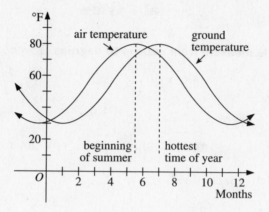

Terms to Know

Example / Illustration

Phase shift (p. 728)

a horizontal translation of a periodic function

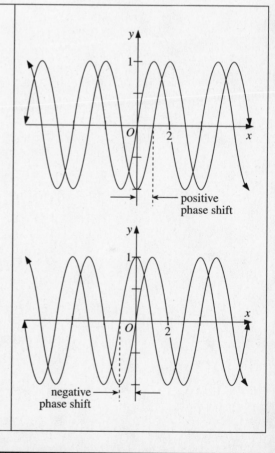

UNDERSTANDING THE MAIN IDEAS

Phase shifts

You can determine the phase shift from a graph by determining the distance between two corresponding points, one on the original graph and the other on the translated graph. In order to determine the phase shift from an equation, the equation must be in the form $y = a \sin b(x - h)$ or $y = a \cos b(x - h)$. The phase shift is $|h|$ units to the right if $h > 0$ and $|h|$ units to the left if $h < 0$.

Example 1

Graph each equation.

a. $y = 4 \sin\left(x - \dfrac{\pi}{2}\right)$

b. $y = \sin\left(2x + \dfrac{4\pi}{3}\right)$

■ Solution ■

Begin by rewriting each equation in the form $y = a \sin b(x - h)$.

a. $y = 4 \sin (1)\left(x - \dfrac{\pi}{2}\right)$

So $a = 4$, $b = 1$, and $h = \dfrac{\pi}{2}$.

The graph is the graph of $y = 4 \sin x$

translated $\dfrac{\pi}{2}$ units to the right.

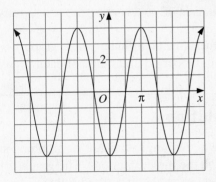

b. $y = \sin\left(2x + \dfrac{4\pi}{3}\right)$

$= \sin 2\left(x + \dfrac{2\pi}{3}\right)$

$= \sin 2\left(x - \left(-\dfrac{2\pi}{3}\right)\right)$

So $a = 1$, $b = 2$, and $h = -\dfrac{2\pi}{3}$.

The graph is the graph of $y = \sin 2x$

translated $\dfrac{2\pi}{3}$ units to the left.

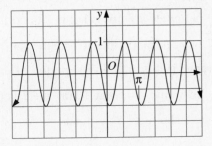

For Exercises 1–6, graph each function. Show at least one period.

1. $y = \cos\left(x + \dfrac{\pi}{2}\right)$

2. $y = 2 \sin 3(x - \pi)$

3. $y = 4 \cos \pi(x - 1)$

4. $y = 3 \sin \dfrac{1}{2}\left(x + \dfrac{\pi}{2}\right)$

5. $y = -5 \cos \dfrac{\pi}{3}(x - 2)$

6. $y = -2 \sin \dfrac{\pi}{2}(x + 3)$

7. Writing Graph the functions $y = \sin\left(x - \dfrac{2\pi}{3}\right)$ and $y = \sin\left(x + \dfrac{4\pi}{3}\right)$ on the same coordinate plane. What do you notice? Make a conjecture about when the graphs of two functions $y = \sin(x + h_1)$ and $y = \sin(x - h_2)$ have a similar relationship.

Combining sine and cosine functions

When you add two periodic functions with different periods you often get a periodic function, but the period of the new function is usually longer than the period of either original function. Also, the graph of the new function may have a more complicated cycle. In general, the sum of two functions $y = f(x)$ and $y = g(x)$ is the function $y = f(x) + g(x)$.

Example 2

In the same coordinate plane, graph $y = \cos x + 2$, $y = \sin \dfrac{2}{3}x + 5$, and their sum. Find the period and range of each function.

■ Solution ■

Use a graphing caluclator or graphing software to graph the three functions. The graphs of the functions are shown at the right.

The period of $y = \cos x + 2$ is 2π and its range is $1 \le y \le 3$. The period of $y = \sin \dfrac{2}{3}x + 5$ is 3π and its range is $4 \le y \le 6$. The period of the sum of the functions, $y = \cos x + \sin \dfrac{2}{3}x + 7$ is 6π and its range is about $5.1 \le y \le 8.9$.

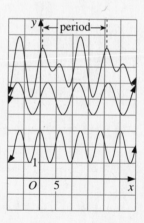

Graph each pair of functions and their sum in the same coordinate plane. Find the period and range of each function.

8. $y = \cos \pi x + 2$

$y = \sin \dfrac{2\pi}{3}x - 5$

9. $y = 2 \sin \dfrac{1}{2}x + 4$

$y = \cos \dfrac{2}{3}x - 4$

10. $y = \sin \dfrac{\pi x}{2} + 7$

$y = 3 \cos \dfrac{\pi}{3}x - 7$

11. Each note produced by a musical instrument consists of a fundamental tone and tones of higher frequencies, called *overtones*. Overtones allow you to distinguish between different instruments, even when both are playing the same note. The physical wave produced by the note B on a clarinet is

modeled by the equation $y = \sin\left(\dfrac{4\pi x}{5}\right) + 2$ (translated upward for clarity),

where the units on the *x*-axis are feet. The lowest overtone of this note that a

clarinet produces is modeled by the equation $y = 0.2 \sin\left(\dfrac{12\pi x}{5}\right) - 2$ (translated

downward for clarity). Graph these two equations and their sum, which models the note produced, in the same coordinate plane.

······················
Spiral Review

Write a sine function with the given maximum *M*, minimum *m*, and period.
(Section 15.3)

12. $M = 1$, $m = -7$, period $= 1.5$ **13.** $M = 8$, $m = 2$, period $= \dfrac{\pi}{3}$

Find the vertical asymptotes of the graph of each function. *(Section 9.8)*

14. $y = \dfrac{6}{(x - 4)(x + 1)}$ **15.** $y = \dfrac{-3}{x^2 - 16}$ **16.** $y = \dfrac{2}{(x + 1)(x + 5)}$

Find tan *A* for each right triangle. *(Section 14.2)*

17.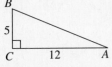

18.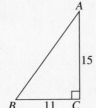

19.

The Tangent Function

Learn how to . . .

- graph the tangent function

So you can . . .

- solve problems involving the tangent function

Application

Transportation engineers describe the steepness of a hill by speaking of the percent of a grade. For example, a 15% grade is a hill that rises 15 ft for every 100 ft of horizontal distance. The percent of a grade is the tangent of the angle of the hill (θ in the diagram), expressed as a percent.

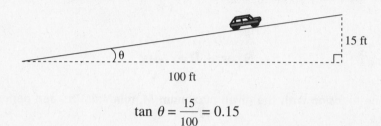

$$\tan \theta = \frac{15}{100} = 0.15$$

Terms to Know

Example / Illustration

Tangent (p. 736)
the ratio of the vertical position of an point on the unit circle to its horizontal position, as a function of the angle its radius makes with the positive x-axis

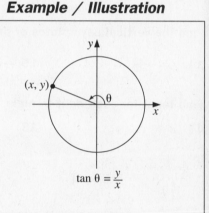

$$\tan \theta = \frac{y}{x}$$

UNDERSTANDING THE MAIN IDEAS

Evaluating the tangent function

You can use a calculator to evaluate a tangent function, or you can use the fact that $\tan x = \dfrac{\sin x}{\cos x}$. Before using your calculator, make sure it is set to the appropriate mode for that function, either degree mode or radian mode.

Example 1

Evaluate each expression.

a. $\tan 40°$

b. $\tan \dfrac{\pi}{3}$

c. $\tan (-3.2)$

Study Guide, ALGEBRA 2: EXPLORATIONS AND APPLICATIONS

■ Solution ■

a. Use a calculator or a table of values. If you use a calculator, make sure it set to degree mode.

$$\tan 40° \approx 0.8391$$

b. You can use a calculator set to radian mode, or you can use the relationship between the tangent function and the sine and cosine functions.

$$\tan \frac{\pi}{3} = \frac{\sin \frac{\pi}{3}}{\cos \frac{\pi}{3}} = \frac{\frac{\sqrt{3}}{2}}{\frac{1}{2}} = \sqrt{3}$$

c. The angle measure is in radians, so set your calculator to radian mode.

$$\tan(-3.2) \approx -0.0585$$

Evaluate each expression.

1. $\tan 30°$

2. $\tan 180°$

3. $\tan \frac{3\pi}{8}$

4. $\tan(-45°)$

5. $\tan \frac{5\pi}{3}$

6. $\tan(-1.4)$

7. $\tan 225°$

8. $\tan\left(-\frac{7\pi}{6}\right)$

The shadow of the outfield wall of a baseball stadium stretches 62 ft beyond the base of the wall when the rays of the sun make an angle of 22° with the ground.

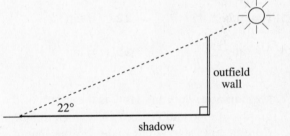

9. How tall is the wall?

10. What happens to the length of the shadow as the sun sets and the angle of the sun's rays decreases?

Graph of the tangent function

Since the tangent function can be written as $\tan x = \frac{\sin x}{\cos x}$, it is undefined whenever $\cos x = 0$, that is, whenever the value of x is an odd multiple of $\frac{\pi}{2}$.

The period of the function $\tan x$ is π. Like the sine and cosine functions, the graph of the tangent function can be stretched vertically or horizontally. The general form of the equation for a tangent function is $y = a \tan b(x - h) + k$, where a, b, h, and k determine vertical and horizontal stretches and shifts.

Example 2

Graph each function. Show at least one period.

a. $y = \tan \frac{1}{2}x$

b. $y = 3 \tan x$

■ Solution ■

a. The graph is a horizontal stretch of the graph of the basic tangent function $y = \tan x$. The period of the graph is 2π, instead of π.

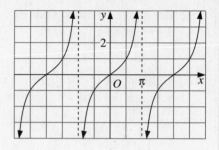

b. The graph is a vertical stretch of the graph of the basic tangent function.

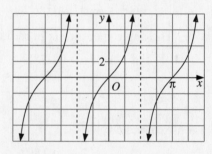

For Exercises 11–16, graph each function. Show at least one period.

11. $y = \dfrac{2}{3}\tan x$

12. $y = \tan\dfrac{\pi}{4}x$

13. $y = -2\tan \pi x$

14. $y = \tan\left(x + \dfrac{\pi}{4}\right)$

15. $y = \tan\dfrac{1}{2}x + 3$

16. $y = 4\tan\dfrac{1}{3}x - 2$

17. Mathematics Journal How is the graph of the tangent function like the graphs of sine and cosine? How does the graph of the tangent function differ from them? Describe as many similar features and as many distinguishing features as you can.

....
Spiral Review

18. Write a sine function that is equivalent to $y = -3\cos x$. *(Section 15.4)*

Evaluate each expression. *(Section 15.1)*

19. $\cos 225°$

20. $\sin(-270°)$

21. $\cos 420°$

22. $\sin 135°$

Chapter 15 Review

Complete these exercises for a review of Chapter 15. If you have difficulty with a particular problem, review the indicated section.

Find all angles θ for $0° \leq \theta \leq 360°$ that satisfy each equation. Give answers to the nearest degree. *(Section 15.1)*

1. $\cos \theta = 0.7880$ **2.** $\sin \theta = -0.9703$ **3.** $\cos \theta = 0$

Evaluate each expression. *(Section 15.1)*

4. $\sin 160°$ **5.** $\sin 245°$ **6.** $\cos (-27°)$

Convert each angle measure from radians to degrees (to the nearest degree). *(Section 15.2)*

7. $-\dfrac{5\pi}{2}$ **8.** $\dfrac{11\pi}{12}$ **9.** -5.1

For Exercises 10–12, convert each angle measure from degrees to radians. Express each as a decimal to the nearest hundredth of a radian, and in terms of π. *(Section 15.2)*

10. $240°$ **11.** $315°$ **12.** $105°$

13. Open-ended Give some advantages and disadvantages of using the radian measure of an angle, rather than its degree measure. *(Section 15.2)*

For Exercises 14 and 15, graph each function. State the amplitude and period of each graph. *(Section 15.3)*

14. $y = 4 \cos \dfrac{\pi}{3}x$ **15.** $y = -\dfrac{3}{2} \sin 4x$

16. Write a sine function with maximum 7, minimum –3, and period 6. *(Section 15.3)*

For Exercises 17 and 18, graph each function. Show at least one period. *(Section 15.4)*

17. $y = 2 \cos \left(x + \dfrac{\pi}{4}\right)$ **18.** $y = -5 \sin \dfrac{\pi}{2}(x - 1)$

19. In the same coordinate plane, graph $y = 2 + \sin \dfrac{2\pi}{3}x$, $y = 4 + \sin \dfrac{\pi}{3}x$, and their sum. Find the period and range of each function. *(Section 15.4)*

For Exercises 20–22, evaluate each expression. *(Section 15.5)*

20. $\tan \dfrac{\pi}{3}$ **21.** $\tan 400°$ **22.** $\tan (-1.6)$

23. Graph the function $y = 2 \tan \left(x - \dfrac{\pi}{4}\right)$. Show at least one period. *(Section 15.5)*

---

24. Writing For what values of θ for $0° \leq \theta \leq 360°$ is $|\tan \theta| > 1$? Explain your answer using a unit circle diagram. *(Section 15.5)*

SPIRAL REVIEW   Chapters 1–15

1. Find the area of the shaded region.

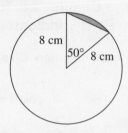

8 cm
50°
8 cm

At a math class picnic, slips of paper with each student's name on them are placed in a hat for a prize drawing. The first student whose name is drawn will win a compact disc. This student's name is *not* placed back in the hat. The second student will win two movie passes. There are 14 girls and 11 boys in the class. Find each probability.

2. P(a boy wins the movie passes $|$ a girl wins the compact disc)

3. P(a girl wins the compact disc and a boy wins the movie passes)

4. P(a girl wins at least one of the two prizes)

An angle θ is in standard position with the given point P on its terminal side. Find sin θ, cos θ, and a value for θ with $0° \leq \theta < 360°$.

5. $P(-12, 35)$ **6.** $P(20, -21)$ **7.** $P(-56, -33)$

Use the binomial theorem to expand each power of a binomial.

8. $(a - 3b)^5$ **9.** $(x^2 + 2y)^4$ **10.** $(10m^3 - n)^6$

Solve each $\triangle PQR$. If there are two solutions, give both. If there is no solution, explain why.

11. $p = 13$, $q = 15$, $r = 19$ **12.** $\angle P = 42°$, $q = 29$, $p = 25$

13. $q = 6.7$, $\angle R = 35°$, $p = 8.4$ **14.** $p = 16$, $\angle Q = 39°$, $\angle R = 105°$